Plant Biology
Science Projects

Best Science Projects for Young Adults

Animal Behavior Science Projects, Nancy Woodard Cain

Plant Biology Science Projects, David R. Hershey

BEST SCIENCE PROJECTS FOR YOUNG ADULTS

Plant Biology
Science Projects

DAVID R. HERSHEY

John Wiley & Sons, Inc.

New York · Chichester · Brisbane · Toronto · Singapore

Copyright © 1995 by David R. Hershey
Published by John Wiley & Sons, Inc.

Library of Congress Cataloging-in-Publication Data:

Hershey, David R.
    Plant biology science projects / David R. Hershey.
        p.   cm.   —(Best science projects for young adults)
    Includes index.
    ISBN 0-471-04983-2 (acid-free paper)
    1. Botany projects—Juvenile literature. [1. Botany—
Experiments. 2. Plants—Experiments. 3. Experiments. 4. Science
projects.] I. Title. II. Series.
QK52.6.H47   1995
581'.078—dc20   94-12934

Printed in the United States of America
10 9 8 7 6 5 4 3 2 1

Contents

How to Use This Book

This book contains about two dozen science projects about seed plants. Seed plants are used because they are readily available, are familiar to students, and are quickly and easily grown indoors. Most of the projects deal with **plant physiology,** which is the study of how plants function, with **plant ecology,** which is the study of a plant's relationships with other organisms and with its environment, and with **plant agriculture,** which is the study of how to grow plants. These three divisions of plant biology were chosen because they provide the greatest number of intriguing science projects that can be quickly, easily, and inexpensively completed by students age 12 and older.

The best way to use this book is to read the Introduction first to get an overview of how to approach a plant biology project. You should then look at the individual projects. There is a step-by-step experiment for each of the projects followed by suggestions for further investigations. Choose a project that interests you, and consider the time, space, and materials required for the project. For most projects, the materials needed are inexpensive and readily available.

Ask your teacher and a parent before making a final choice of a project. Show them the project directions, and explain what you will need. Some projects need adult help in buying supplies, building equipment, or working with potentially dangerous tools and chemicals. Be sure that your teacher or other responsible adult agrees to provide the help you need. Also, discuss where you will do the project.

You will have to refer to other sections of the book because details common to many projects are given in the Introduction, appendices, and the introductory section of each of the five groups of projects. For example, discussion of desirable plants for projects and examples of graphing data are found in the Introduction. Construction of fluorescent light systems, uses for 2-liter soda bottles, and suppliers are given in the appendices. Books and articles for further reading are given at the end of each introductory section to a part and at the end of some projects.

Introduction

WHY PLANT EXPERIMENTS?

People experiment with plants for two main reasons. First, plants are fundamental to our survival. Either directly or indirectly, they provide us with many essential products, including virtually all of our food, clothing, oxygen, paper, fuels, plastics, and much of our medicine and shelter. Other important plant products are perfumes, soaps, dyes, inks, paints, rubber, rope, alcohol, sports equipment, musical instruments, and photographic film. Plants enhance our environment by absorbing carbon dioxide, providing shade, reducing soil erosion, preventing flooding, reducing noise, serving as windbreaks, and cooling the air. Plants also provide settings for recreation and surround us with beauty, through their flowers, fragrances, and colorful autumn leaves.

The second important reason we experiment with plants is because we are curious about these organisms that are alive but that are so different from human beings. How does a plant's life differ from yours? How is it the same?

If you're looking for a biology project, plants offer several advantages. They are inexpensive, readily available, and adaptable to school and home environments. Unlike animals, plants do not bite, run away, or have to be cleaned up after. Also, you can treat plants in ways that you cannot humanely treat animals. The great diversity of plants makes an unlimited number of experiments possible. Plant projects also help you develop gardening skills that you can use to grow better houseplants, lawns, vegetables, trees, and shrubs. Extra plants from your project could be used as gifts, eaten, sold, or planted in your garden.

BASIC TYPES OF EXPERIMENTS

Most plant science projects examine the effect of one or more experimental treatments on one or more types of plants. The effect of the treatment is determined by measuring the plants. There are hundreds of choices for the experimental treatments including light intensity, light color, hours of light per day, temperature, humidity, chemical concentration, fertilizer, irrigation method, irrigation frequency, soil type, soil pH, soil compaction, and soil volume, just to name a few. Plants can also be measured in many different ways including stem height, leaf number, flower number, mass, and leaf area.

Science projects with plants can be of several basic types, of increasing complexity.

1. The simplest type of project is the "plus-or-minus treatment" study. An example of this type would be seed germination in plus-light or minus-light.

2. An expansion of the plus-or-minus treatment study is the "dose–response" study, which uses three or more levels of a treatment factor—for example, 0, 1, 2, 3, 4, and 5 g of fertilizer per plant. The data in the dose–response study are easily graphed with the response on the vertical axis and the dose on the horizontal axis (Figure I.1).

3. The "factorial experiment" is a dose–response study using two or more treatments, for example, four rates each of fertilizer and pesticide. The preceding example would be a 4-by-4 **factorial,** and there would be sixteen separate treatments.

4. Another type is the "comparative experiment," which compares two or more types of plants, each grown under the same conditions. It can also compare the application of several substances, such as fertilizers, each at one rate.

SCIENTIFIC NAMES

Scientists in their reports always give the scientific name of the plant(s) they experiment with and you should too. It is only necessary to list the scientific name once in a project, then you can use the common name of the plant the rest of the time. Often, the scientific name can be placed in

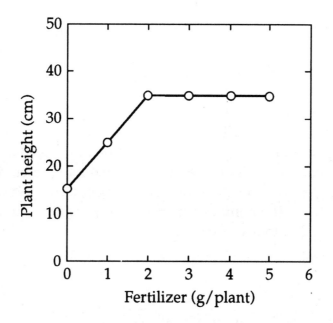

Figure I.1. A typical dose–response graph showing dose (in this case, fertilizer) on the horizontal axis and response (in this case, plant height) on the vertical axis.

parentheses after the common name in the title—for example, "**Phototropism** of Corn (*Zea mays*)." Always capitalize the first, or **genus,** name and underline (or italicize) both names. Note that *Zea mays* is the **species** name. *Zea* is the genus, and *mays* is the **specific epithet,** not the species. Many common plants are **cultivars** (cultivated varieties), whose names are enclosed in single quotation marks after the scientific name—for example, tomato (*Lycopersicon esculentum* 'Big Boy'). The scientific names of most plants you might work with are found in the following reference:

Bailey, L. H., E. Z. Bailey, and Bailey Hortorium Staff. 1976. *Hortus third: A concise dictionary of plants cultivated in the United States and Canada.* New York: Macmillan.

STANDARDS IN PLANT EXPERIMENTS

Control and Replication

All experiments should have a **control,** which is usually a plant that receives a treatment level of zero or a normal level. For example, in a fertilizer dose–response experiment, the control would receive no fertil-

izer, and the other plants would each receive a different dose of fertilizer. In a nutrient-deficiency study, the control would receive a normal
level of fertilizer while the other plants would receive different doses
that are less than the normal level.

Scientists always have more than one experimental unit, or **replication,** per treatment. A replication is usually one plant. In a seed experiment, a replication may be a batch of 20 or more seeds. Replication
accounts for the many differences that exist among different individuals
of the same species. Consider how different people are, yet we are all of
the same species. The minimum number of replications should be three
per treatment, but four or five is even better.

Compiling Data

For each project, you will need to define what factors you will be measuring. Choices include plant height, color, mass, leaf area, root length,
numbers of leaves, flowers or fruits, chlorophyll content, starch content,
and so on. The factors you choose will depend on what questions your
experiment is trying to answer. It is best to measure as many different

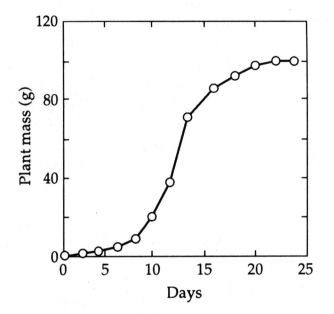

Figure I.2. A time-course graph showing the typical S-shaped pattern of plant mass
increase with time.

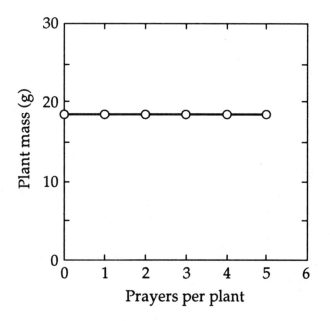

Figure I.3. A dose–response graph in which the treatment factor (in this case, prayers per plant) has no effect on plant mass.

factors as possible, because the more data you have the more questions you will be able to answer.

At a minimum, you should measure each factor at the start and at the end of the experiment. It is even better to take measurements several times during the study. This allows you to create a **time-course graph** (Figure I.2), which shows how a factor changed over a certain period. The period can be in hours, days, weeks, or whatever is most logical for your experiment. A time-course graph for plant growth often gives an S-shaped curve. The growth increases slowly at first. Then, there is a very fast period of growth. Finally, the growth slows and may stop, as in the figure.

Plant Responses

Most plant projects examine the response of plants to two or more levels of an environmental factor, such as light, fertilizer, water, or chemicals. Basic plant responses are of several general types. If the factor has no effect on the plant, the graph is a horizontal line (Figure I.3). If the factor is "toxic," the curve will slope downward (Figure I.4).

If the factor is essential to the plant, the line will bend upward first as the factor moves up in the deficient range of concentration (Figure I.5).

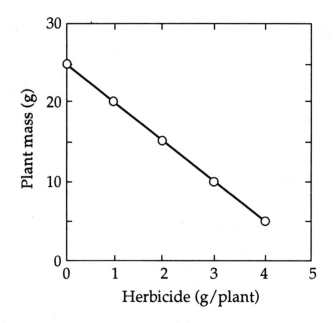

Figure I.4. A dose–response graph in which the treatment factor (in this case, herbicide concentration) has a toxic effect on plant mass.

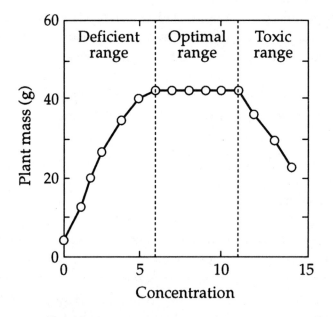

Figure I.5. A dose–response graph in which the treatment factor is essential for the plant. As the treatment concentration increases, the plant mass increases in the deficient range, stabilizes in the optimal range, and finally decreases in the toxic range.

The plant does not have enough of the factor in the deficient range, so growth increases as the concentration increases. Then, the line will become flat when the factor is in the optimal range of concentration. The plant has all of the factor it needs in the optimal range, so increasing the concentration does not increase growth. Finally, the line will go down when the concentration is in the toxic range because the plant has too much of the factor.

Metric System

All measurements in this book are given in metric units because scientists always use metric units. The metric system is also the official system for all measurements in the United States and virtually all other nations. Unfortunately, the United States still uses mostly nonmetric units, such as inches, feet, quarts, gallons, and degrees Fahrenheit. Thus, you will often have to convert nonmetric to metric units and must be familiar with two systems of measurement.

In this book, length is given in centimeters (cm) and volume in milliliters (ml) or liters. Scientists say mass instead of weight and express mass in grams (g) or kilograms (kg). Temperature is expressed in degrees Celsius (°C). Concentration is typically expressed as milligrams/liter

TABLE I.1   Changing Nonmetric to Metric Units

Nonmetric unit	×	Conversion factor	=	Metric unit
Length				
Inches		2.54		cm
Feet		30.5		cm
Mass				
Ounces		28.4		g
Pounds		454		g
Volume				
Teaspoons		4.9		ml
Tablespoons		14.8		ml
Fluid ounces		29.6		ml
Cups		236.5		ml
Quarts		946		ml
Gallons		3.78		liters

Temperature
[Degrees Fahrenheit (°F) − 32] × 5 ÷ 9 = degrees Celsius (°C)

(mg/liter). Note that metric abbreviations are not followed by a period while nonmetric units are—for example, pound (lb.) and gram (g).

Try to make all your measurements in metric units. If you must make measurements in nonmetric units, use the conversion factors in Table I.1 to convert your measurements into metric units. Multiply the nonmetric unit in the first column of Table I.1 by the conversion factor in the second column to get the metric unit in the third column. For example, 5 inches multiplied by 2.54 equals 12.7 cm. Temperatures require several more steps for conversion. Subtract 32 from degrees Fahrenheit, then multiply by 5, and then divide by 9. That gives degrees Celsius—for example, $212°F - 32 = 180 \times 5 = 900 \div 9 = 100°C$.

WHAT YOU WILL NEED

Instruments

The most common instruments used in plant science projects are rulers, graduated cylinders, and balances. Biological supply companies sell these and everything else you will need for a plant science project (see Appendix 5). Plastic graduated cylinders are more desirable than glass ones because they do not break. The 100-ml size is probably the most versatile. Kitchen measuring cups and spoons can be used if graduated cylinders are not available. However, measurements will have to be converted to metric units. One- or 2-liter plastic soda bottles are convenient for measuring water for nutrient solutions.

Electronic balances have come down in price, with many retailing for less than $100–150, and mechanical triple-beam balances retail for just under $100. Most schools have balances that you should be able to use. If your school does not have balances, have your teacher ask the police department if they have any confiscated balances that they will donate to the school. You may be able to use postal scales for some measurements.

The **pH** of a solution or soil sample is often measured in plant projects. The pH is a measure of how acidic a substance is. Do not use metal-probe pH meters for this; they are extremely inaccurate. You can use inexpensive pH papers or test kits, but they are typically accurate to only 0.5 units. If you can, get a pocket-size glass electrode pH meter. They cost as little as $30 and are accurate to 0.2 units. You can also get test kits for many other factors of interest including nutrient levels in soil, water, and plant tissue.

Another common instrument owned by most schools and many homes is the personal computer. Word-processing software can be used to prepare your project report and print the large lettering for the poster. Spreadsheet software is excellent for statistical analysis of the data and for creating graphs. Drawing programs can be used to create professional-quality line drawings of experimental setups. Computers connected to networks can also be used to search scientific literature databases, such as BIOSIS, AGRICOLA, and ERIC.

One of the most underutilized instruments is the camera. A 35-mm camera can be used to illustrate experimental setups and treatment effects on plant size, shape, and color. A video camera can also be very useful to record project results. Many high schools have a spectrophotometer, which can be used to quantify the amount of chlorophyll, carbohydrates, protein, and many essential mineral nutrients in plant tissue.

Equipment

Much of the equipment needed for plant projects can be made from recycled plastic containers, including 1- and 2-liter soda bottles (see Appendix 1), 35-mm film cans, yogurt cups, and margarine tubs. Disposable plastic Petri dishes are very useful in seed germination studies but, if necessary, can be replaced with plastic picnic plates enclosed in a plastic bag. Filter paper circles can be replaced with coffee filters or circles cut from paper towels.

Light

Plant growth needs light energy, so it is not surprising that lack of light is one of the biggest limitations faced in plant science projects, especially because science fair projects are often done in the winter when sunlight is low. Fortunately, light limitations can be overcome by using fluorescent lights. A 120-cm-long fluorescent shop fixture with two 40-watt cool-white tubes costs about $8–10. Even better is a light bank with three or four of these shop fixtures (see Appendix 2). A miniature plant growth chamber can be easily assembled using a 30-watt circular fluorescent bulb, which screws into a standard incandescent fixture, and a 19-liter (5-gallon) plastic bucket (see Appendix 2).

If you do not have a fluorescent light system, choose plants that require low light levels, allow extra time for plant growth in winter, or do your experiments in spring and summer. Another option is to do projects that do not require high light for plant growth, such as many seed projects.

Soil, Water, and Fertilizer

Problems with soil, water, or fertilizer often limit plant growth in pots. A common mistake is to use soil from your yard or garden. Garden soil usually stays too wet when put in a pot and often contains weed seeds, diseases, and insects. Instead, buy **potting soil,** which has been heat-treated to kill weed seeds, insects, and diseases. Another common error is to tightly pack the soil around the roots. Tight packing eliminates air channels in the soil. This harms roots because they require oxygen. When potting a plant, firm the soil gently by tapping the pot several times on a flat surface. Watering the pot will further settle the soil.

Potting soils typically do not contain field soil but are a mixture of two or more materials, such as **perlite, vermiculite,** and **sphagnum peat moss.** Perlite and vermiculite are minerals that are heated to about 1000°C, which causes them to expand like popcorn. They are very light-weight because they contain a lot of air, but are easily crushed. Perlite or vermiculite is also often used by itself to **root** cuttings, to germinate seeds, or in place of potting soil for experimental plants because it is more easily removed from the roots than potting soil. Sphagnum peat moss is usually mixed with another material(s) to make a potting soil. It consists of partially decomposed moss plants and is obtained from bogs in Canada and the northern United States.

Overwatering is another common cause of plant failure. Overwatering can be avoided by having a drainage hole in the bottom of every pot, using a potting soil, and not packing it tightly. In winter, you should let water come to room temperature before irrigating the plants to avoid cold soil temperatures, which slow plant growth.

Houseplant fertilizers usually work well if you follow their directions. Avoid the tendency to give plants more than the recommended rate. It almost always does more harm than good. Place a saucer beneath each pot so that excess water can drain from the pot and flush away excess fertilizer. It will also protect your furniture from water damage.

PLANNING YOUR PROJECT

Most projects can be done at home or in the school classroom with minimal equipment. For more advanced experiments, several of the projects could be combined. Directions are for one replication only; however, you should try to have a minimum of three replications. For example, in the deficiency study (see Chapter 20) you should try to have at least three control plants that get a normal level of mineral nutrients, three plants that do not get any calcium, three plants that do not get any iron, and so on.

Project Location

Where the project will be conducted is an important consideration. If space is available at school, there are several advantages to doing the project there. For example, your teacher may be available to help supervise the project, you will have better access to scientific instruments, and family members or pets cannot accidentally ruin the project. Possible disadvantages of school-based projects include vandalism by other students and the difficulty of gaining access to the project after school or on weekends. You also need to find out if the heat is turned back or the electricity shut off after school hours. If they are, you should not leave your project there. Cool temperatures slow plant growth, and fluorescent lights should be run continuously in most plant growth experiments.

If you do a project at home, try to find a place where your experiment will not be disturbed while plants grow or seeds sprout. Your bedroom or a spare room might be best. Take your pets and younger brothers and sisters into consideration. You don't want them to ruin your project. You also do not want them to be hurt if they get into your experimental supplies or equipment. You can work on the parts of your project that use a sink or the stove in the kitchen. If you will be using windows as the light source, the brightest window may dictate the project location.

Potting soil preparation and potting should be done in a garage or basement. You'll need a heated area in cold weather. To minimize dirt, place several layers of newspaper on your work surface. When you are finished working with the potting soil, scatter soil clinging to the newspapers outdoors around shrubs or on a lawn. Then, refold the newspapers and return them to the recycling bin.

Record Keeping

Complete and accurate records are very important in science projects. Obtain a notebook to record all information and data about your project. Write down all your observations and measurements as you make them. Do not rely on your memory. Also, always write down the date and time of day when you make observations and measurements.

Labeling and Pots

All experimental plants should be labeled clearly with a waterproof marking pen or pencil. Do not try to just remember which plant is which. You can buy plastic or wood labels to stick in the soil of plant containers, or you can make labels out of wood popsicle sticks or cut from plastic containers.

A wide variety of containers can be used as plant pots. Plastic containers are preferable because they are inexpensive, lightweight, easily washed, widely available, and less breakable than glass or clay pots. Cardboard or paper containers tend to rot, and metal containers tend to rust. Many types of plastic pots and flats are sold specifically for plants; however, many other plastic containers can be recycled into plant pots. These include soda bottles, plastic cups, yogurt cups, margarine tubs, and Styrofoam™ egg cartons. All plant containers that use potting soil should have one or more drainage holes in the base of the container. With adult help, cut drainage holes with an electric drill or knife, or punch holes with a hammer and nail.

Project Presentation

A written project report is usually required. You will also have to prepare a poster if your project is entered in a science fair. Be sure to prepare your report and poster carefully because their quality is important in grading and judging. A well-done report or poster of an ordinary project often makes a better impression than a poorly done report or poster of an extraordinary project. For a professional appearance, print the report and poster text, graphs, and tables using a computer printer.

You can also prepare an attractive poster without a computer using stencils and lettering guides. Black text on white paper is best. Add color by using colored poster board as a background for the text and illustrations. Be sure to follow any instructions your teacher gives you on poster preparation. A report and poster typically have the following parts: title, abstract (summary), introduction, procedures, results, discussion, and references.

Title

A common error is to include unnecessary words in the title. Phrases like "Effect of," "Measurement of," "The Study of," "A Comparative Study of," and "Different Kinds of" are not necessary. A possible title for the project in Chapter 1 is "Salt and Bean Seed Imbibition." This is a shorter title than "A Study on the Effect of Various Concentrations of Salt on Bean Seed Imbibition." Both titles, however, give the same information. We know that science projects usually look at the effect of two or more levels of a factor on some aspect of plant growth. Thus, the title does not need to include things that we take for granted.

A good title has a minimum of three parts: the plant name, the factor(s) measured, and the treatment. In the preceding example, the plant name is bean, the factor studied is imbibition, and the treatment is salt. You may include the plant scientific name in parentheses after the common name—for example, "Bean Seed (*Phaseolus vulgaris*)." Include the common name because most of your audience is unfamiliar with scientific names.

Do not make the title too big or you will not have enough space for the other parts of the poster. Use all capital letters for the title. I make the title letters in my posters 2 cm tall.

Abstract and Introduction

The abstract is a brief summary of the project. It is typically 150 words or less. The abstract should be written after the rest of the report is done. The introduction provides the reader some basic information about the topic and tells the purpose or objective of the experiment. It may also tell the **hypothesis,** which is your educated guess of what the experimental results will be.

Procedures

The procedures section briefly states what materials you used and what you did in the experiment. Scientists often call this the "Materials and Methods" section. It will usually not be as detailed on the poster as in the written report for two reasons. One is that there is not enough space on the poster. Also, you usually can display your experimental equipment at the science fair. The audience can see what you did rather than read about it.

For example, the poster might say,

> Bean (*Phaseolus vulgaris* 'Bush Baby') seeds were sown in 400-ml plastic pots. Ten days after sowing, they were sprayed with sulfuric acid solutions with a pH of 3, 4, 5, or 6. Plant height and leaf area were measured 7 days after spraying.

This gives the key experimental methods; however, the written report would include much more detail including the type of soil, how you measured the pH and leaf area, the light source, the temperature, how the plants were irrigated, and whether or not the plants were fertilized.

Results and Discussion

In the results section, you tell what happened in the experiment. Most of your results can be given in tables, graphs, and illustrations, such as photos, drawings, or preserved plants. In the discussion section, you try to explain why you got the results you did and what you learned from them. You will also discuss how your results either agree or disagree with your hypothesis. Scientists often combine the results and discussion into one section to make it easier to refer to the results.

References

A reference section lists all the books and articles that gave you information you used in the project. You can also list scientists or other people who gave you information. Usually, you will only include the references in the written report because there is not enough space on the poster.

EXPERIMENTAL PLANT SPECIES

One of the great attractions of plants is their huge variety. There are over 250,000 species of flowering plants. Adding even more diversity are the cultivated plants, termed cultivars, which is short for "cultivated variety." There are several hundred thousand plant cultivars including 10,000 wheat cultivars and 5000 rose cultivars.

For science projects, the choice of plants is more limited for several reasons including availability, cost, size, light requirements, and speed of growth. Available and inexpensive plants include supermarket plants, houseplants, and flower, vegetable, weed, and lawn grass seed. Space is often limited for projects, so small or dwarf species are desirable. Light is often limited in homes and classrooms. Thus, houseplants that tolerate low light levels are an advantage. Rapid growth is usually desirable so the project can be completed quickly.

Wisconsin Fast Plant

Wisconsin fast plant (*Brassica rapa*) is considered a model plant for teaching because of its 15–20 cm height, five-week life cycle, two-week time from seed to flower, and availability of several dozen cultivars. Wisconsin fast plants are especially useful for genetic studies because they are easily **hand-pollinated. Pollination** is the transfer of **pollen** to the **stigma** of the flower. Pollination is often done by wind, insects, or bats. When done deliberately by a person, it is called hand-pollination. Disadvantages include their requirement for a fluorescent light system, relatively high seed cost, and need for **cross-pollination** to obtain seed set. Cross-pollination means that pollen from the **anther** of one plant must be transferred to the stigma of another plant. Wisconsin fast plants are sold by Carolina Biological Supply (see Appendix 5).

Devil's Backbone

Devil's backbone or bryophyllum (*Kalanchoe daigremontiana*) is a common houseplant that produces numerous plantlets on the edges of its leaves. This makes it easy to propagate large numbers of identical plants. Devil's backbone also needs infrequent watering, can be grafted, has crassulacean acid metabolism (see Chapter 13), and is ideal for **photoperiodism** studies.

Houseplants

Many houseplants are excellent for projects because they are readily available, easily rooted, and tolerant of low light. Wandering Jew (*Zebrina pendula*), giant white inch plant (*Tradescantia albiflora* 'Albovitata'), heartleaf philodendron (*Philodendron scandens* subspecies *oxycardium*), pothos (*Epipremnum aureum*), piggyback plant (*Tolmiea menziesii*), coleus (*Coleus × hybridus*), and geranium (*Pelargonium × hortorum*) are some of the best for science projects. The first four in the list are vines, which means they do not grow too tall.

Piggyback plant is one of the few houseplants native to the Pacific Northwest. It is a **rosette** plant, meaning that the stem is very short and the leaves form a compact cluster. The rosette form is an advantage because the piggyback plant does not get very tall. The piggyback plant forms a plantlet at the base of its leaf. The leaf can be removed from the plant and rooted to form a new plant.

Coleus and geranium have long been used in teaching. A green and white cultivar of coleus is a standard plant for starch testing (see Chapter 11). Geranium is also often used in starch testing as well as in transpiration experiments (see Chapter 7).

Supermarket Plants

Supermarket plants include pineapple tops, carrot, sweet potato, potato, onion, avocado pits, citrus seeds, dry beans, and popcorn. They are readily available and easily grown in a classroom. However, some grow too slowly for science projects, such as pineapple and citrus. Others, such as carrot and sweet potato, tend to get too big. The most useful are dry beans, which can be used in some seed experiments.

Novelty Plants

Novelty plants include chia (*Salvia hispanica*), sensitive plant (*Mimosa pudica*), Venus flytrap (*Dionaea muscipula*), peanut, cotton, redwood burl, ti plant, resurrection plant, voodoo lily, and baobob. The most useful of these for science projects are the first three. Sensitive plant is noted for its leaves that fold when touched. The Venus flytrap is a carnivorous plant that traps and digests insects and small animals.

Chia, of Chia Pet® fame, was a major food crop of some southwestern American Indians and the Aztecs. Its seeds are rich in oil and protein. They were mixed with water and eaten. The seed is unusual because it has a jellylike seed coat. The sticky seed coat makes the seed ideal to stick to the ceramic plants or Chia Pets. Its interesting history and unusual seed may give your science project a special appeal.

Chia is a good substitute for Wisconsin fast plant in experiments that last only one to two weeks and do not require flowers. Chia seeds are much cheaper than Wisconsin fast plants and germinate rapidly. However, chia eventually gets too tall to grow under fluorescent lights.

Flower and Vegetable Seeds

Flower and vegetable seeds are excellent for science projects. They are inexpensive, readily available, and grow rapidly under fluorescent light systems (see Appendix 2). Vegetable seeds useful for projects include lettuce (*Lactuca sativa*), radish (*Raphanus sativus*), bush bean (*Phaseolus vulgaris*), corn, and dwarf tomatoes. Lettuce and radish can be ready for harvest in three to four weeks under a fluorescent light system (see Appendix 2). Lettuce 'Salad Bowl' and 'Grand Rapids' are excellent under fluorescent light systems. Generally, corn, bean, and tomato get too big or take too long to mature for most indoor situations; however, they are useful for short-term experiments using seedlings.

Useful flower seeds include the following dwarf cultivars: 'Jewel Box' cockscomb (*Celosia cristata*), 'Red Poncho' coleus (*Coleus* × *hybridus*), 'Mini-Star' gazania (*Gazania rigens*), 'Super Elfin' impatiens (*Impatiens wallerana*), 'Bolero' marigold (*Tagetes patula*), 'Pink Splash' polka-dot plant (*Hypoestes phyllostachya*), 'Little Pinkie' periwinkle (*Catharanthus roseus*), and 'Thumbelina' zinnia (*Zinnia elegans*). I have grown all these successfully under a fluorescent light bank system. The zinnia and marigold were in bloom six weeks after sowing. The celosia, gazania, and impatiens were in bloom eight weeks after sowing. Polka-dot plant grew much better with 12 hours of light per day than with continuous light.

FURTHER READING

Graf, A. B. 1985. *Exotica, series 4 international: Pictorial cyclopedia of exotic plants from tropical and near-tropical regions: A treasury of indoor ornamentals for home, the office, or greenhouse.* East Rutherford, N.J.: Rhoers.

Hessayon, D. G. 1993. *The house plant expert.* London: Expert Books.

Hibbs, E. T., and N. G. Yokum. 1976. Bryophyllum: A versatile plant for the laboratory. *American Biology Teacher* 38: 281–83.

Huxley, A. 1985. *Green inheritance: The World Wildlife Fund book of plants.* Garden City, N.Y.: Anchor Press.

Lewington, A. 1990. *Plants for people.* New York: Oxford Univ. Press.

Williams, P. H. 1989. *Wisconsin fast plant manual.* Burlington, N.C.: Carolina Biological Supply Company.

Wyman, D. 1986. *Wyman's gardening encyclopedia.* New York: Macmillan.

PART I

Seeds

Seeds are excellent for science projects because they are inexpensive and readily available and take up limited space, which allows for adequate replication. In seed experiments, each replication often consists of 20 or more seeds per treatment. Most seed experiments do not require as much light as projects that involve the growing of plants. Also, seed experiments are usually quick, so they can be repeated several times. They also make good last-minute projects, not that you should do your project at the last minute. However, if a more challenging project does not work as planned, a seed project could be completed quickly as a substitute project.

Seeds vary greatly in size and shape. The largest seed is the double coconut, with a mass of up to 30 kg. Orchids have dustlike seeds with up to 4 million seeds per capsule. Petunia seeds (7000 seeds per gram) and wax begonia seeds (70,000 seeds per gram) are common seeds that are also very small.

University of Wisconsin Professor Paul Williams has defined a seed as "a baby plant in a box with its lunch." This is a nice simple definition, but scientists have many specialized terms to describe the parts of seeds. Scientists call the baby plant an **embryo.** The embryo consists of the baby root called the **radicle,** the baby shoot called the **plumule,** and the baby leaf called the **cotyledon.** The **epicotyl** is the embryonic stem above the cotyledon(s). The **hypocotyl** is the embryonic stem below the cotyledon(s).

Only plants classified as **gymnosperms** and **angiosperms** have seeds. Gymnosperm seeds have a variable number of cotyledons, usually more than two. Angiosperm seeds that have only one cotyledon, such as corn, rice, onion, and grasses, are called **monocots,** or **monocotyledons.** Angiosperm seeds that have two cotyledons, such as bean, radish, tomato, and lettuce, are called **dicots,** or **dicotyledons.** Ferns, mosses, and other primitive plants have spores, but not seeds.

The "box" enclosing the baby plant is called the **seed coat,** which has a small scar, called the **hilum,** where it was attached to the parent plant. Some seeds also have a **micropyle,** a small pore in the seed coat near the hilum. The micropyle is where the pollen entered the seed. The "lunch" is contained in the enlarged cotyledons of some seeds. In some other types of seeds, such as corn and wheat, the lunch is found outside the embryo but within the seed coat in a tissue called the **endosperm.**

A common example of a dicot seed is the bean (Figure PI.1). The outside of the seed coat has a hilum with a small micropyle to one side. The bean seed coat is easy to remove after soaking the seed. Most of the bean seed consists of the two cotyledons attached to the rest of the embryo lying between them. The plumule and radicle are visible. The epicotyl and hypocotyl are not really apparent. They do not grow longer or elongate until the seed sprouts, or **germinates.** When a bean seed germinates, the radicle and hypocotyl elongate first. The cotyledons are carried aboveground by the elongating hypocotyl. The epicotyl then elongates.

A common example of a monocot "seed" is corn (Figure PI.2). What we call a corn grain is not just a seed; it is actually a **fruit.** A fruit is a ripened **ovary,** which includes the seed and the tissue surrounding the seed. The corn seed coat is fused to the ovary wall on the outside and to the other seed tissues on the inside. Therefore, the corn seed coat is not easily removed.

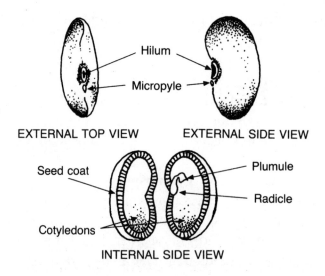

Figure PI.1. External and internal structures of a bean seed, a dicot.

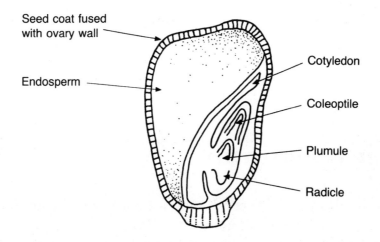

Figure PI.2. Internal structures of a corn seed, a monocot.

If you study seeds, you should also consider the ways that people use seeds. Many of our foods are seeds, such as corn, rice, wheat, oats, barley, rye, millet, beans, peas, soybeans, peanuts, and popcorn. We even decorate our breads and rolls with caraway, poppy, and sesame seeds. Nuts are seeds, including almonds, cashews, chestnuts, filberts, macadamias, pecans, pistachios, and walnuts. Some spices are seeds, like nutmeg and aniseed. Butter and margarine are colored with annatto, a dye made from the seeds of an Amazon tree, *Bixa orellana*.

Our coffee, chocolate, and cola drinks come from seeds. Sweeteners from corn seeds (corn syrup) are widely used in many food products. Much of our alcohol is fermented from seeds. Some of this alcohol is being mixed with gasoline to fuel our cars. Most cooking oils are pressed from seeds including coconut, corn, rapeseed, safflower, soybean, and sunflower oils. Linseed and tung seed oils are major ingredients of paint. Jojoba seed oil is used as a substitute for sperm whale oil in lubrication, an example of a seed helping to save an endangered animal. Jojoba seed oil is also used widely in cosmetics. Sick children used to be given doses of castor seed oil. Today, castor oil is an important ingredient in lipstick and bath oils.

You probably do not realize that a lot of our clothing comes from seeds. Cotton fibers are part of the cotton seed coat. The shiny red and black seeds of rosary pea (*Abrus precatorius*) are used in jewelry but are poisonous if eaten. Indian corn is a popular decoration for Halloween and Thanksgiving. Popcorn strings are common Christmas tree decorations.

Seeds are even used as toys, such as Mexican jumping beans. In the English game conkers, a hole is drilled in a buckeye seed, and a knotted string is put through the hole. The players then take turns swinging the buckeye on a string at another buckeye. The first buckeye to break is the loser.

FURTHER INFORMATION

Write for seed catalogs (see Appendix 5). Visit a state seed testing laboratory or a state forest tree seedling nursery. Contact local greenhouses, plant nurseries, and farms, all of which use seeds. Unusual tree and shrub seeds may be available at botanical gardens and arboretums. Always ask the landowner for permission before you collect any plants or plant parts, including seeds.

FURTHER READING

Hartmann, H. T., and D. E. Kester. 1983. *Plant propagation: Principles and practices.* Englewood Cliffs, N.J.: Prentice-Hall.

Reilly, A. 1978. *Park's success with seeds.* Greenwood, S.C.: George W. Park Seed Company.

U.S. Department of Agriculture. 1974. *Seeds of woody plants in the United States.* USDA Handbook 450. Washington, D.C.: U.S. Government Printing Office.

CHAPTER 1

Bean Seed Imbibition

The process of dry seeds absorbing water and swelling is called **imbibi-tion,** or **hydration.** Seed imbibition is the first step in seed germination, or sprouting. It is also common in cooking when dry rice, beans, or peas absorb water and increase in volume two to three times. Imbibition also occurs in wood doors and door frames during humid weather, causing doors to stick. Another example of imbibition is the swelling of a dry sponge when it absorbs water.

In this project, you will examine the time course for seed imbibition and the pressure generated by imbibing seeds.

TIME COURSE

Method I. Graduated Cylinder

Purpose

To measure the rate of seed volume increase during imbibition.

Materials

☐ 30 ml dry seeds, such as navy bean (available at a supermarket)
☐ 100-ml graduated cylinder
☐ water

Procedure

1. Place 30 ml of dry seeds in the 100-ml graduated cylinder and fill with water to the 90-ml line. If the water surface is slightly curved in

the graduated cylinder, use the bottom of the curved water surface as the measuring point for determining the water volume.

2. Record the volume of seeds by checking what line they come up to in the graduated cylinder. Do this at the beginning of the experiment and again every hour for 12 hours or until the volume no longer changes (whichever comes first).

3. Also every hour, record the volume of water in the graduated cylinder.

4. Graph the seed volume in milliliters against the time in hours from 0 to 12.

Expected Results

The seed volume increases with time as the seeds imbibe water. Eventually the seed volume stops increasing as imbibition is complete.

Why?

Water moves into a dry seed due to **adhesion,** which is the attraction between unlike molecules. The water molecules are attracted to the molecules in the dry seed. This is the same process that occurs when water is absorbed by a dry sponge or a dry paper towel. The dry seed swells as water is absorbed and increases in volume. The seed swelling helps the seed to break the seed coat, which is needed so the plumule and radicle can emerge.

Further Investigations

1. Is imbibition complete after 24 hours of soaking? If not, try to determine how long it takes.

2. What is the pattern of seed volume increase? A straight line indicates that the seed volume increased at a constant rate. A curved line indicates that the rate of swelling was changing.

3. Does removing or cutting the seed coat affect the rate of imbibition?

4. Did the water level in the graduated cylinder change from the initial value? Try to explain why it should or should not change.

Method II. Balance

Purpose

To measure the rate of seed mass increase during imbibition.

Materials

- ☐ balance
- ☐ 240 g dry seeds, such as navy bean
- ☐ 12 plastic cups
- ☐ water
- ☐ paper towels

Procedure

1. Place 20 g of dry seed in a plastic cup and cover with 3 cm of water.
2. Repeat step 1 with a new 20-g batch of seeds in a new cup every hour for 11 hours.
3. At hour 12, pour off the water, briefly blot each batch of seeds with paper towels to remove unabsorbed water, and weigh each batch of seeds.
4. Calculate how much water the seeds imbibed as a percentage of the dry mass. To do this, use the following formula: (final seed mass − 20 g) ÷ 20 g × 100. The final seed mass includes imbibed water and dry seed. Subtracting the 20-g mass of dry seed from the final seed mass gives the mass of water imbibed. Dividing the mass of water imbibed by the 20-g mass of dry seed, and multiplying by 100 gives the percentage of water absorbed by the dry seed. For example, if the final seed mass is 50 g, the percentage of increase is 150.
5. Graph the percentage of seed water content for each batch of seeds against the time each batch soaked in water, in hours.
6. Graph the mass of each batch of seeds against the time each batch soaked in water, in hours.

Expected Results

Both the percentage of seed water content and seed mass should increase over time as the seeds imbibe water.

Why?

Both the mass and volume of seeds increase as the seeds absorb water. One milliliter of water has a mass of 1 g, so the volume and mass increases should be similar. Measuring imbibition by the increase in mass is more accurate than measuring the increase in volume. This is because there are many spaces between the seeds that also occupy volume.

IMBIBITION PRESSURE

Purpose

To determine how much pressure imbibing seeds can generate.

Materials

- ☐ fifteen 35-mm film cans
- ☐ dry seeds, such as navy bean
- ☐ water
- ☐ balance (optional)
- ☐ table salt (optional)

Procedure

1. Fill five 35-mm film cans level at the top with dry seeds and slowly fill with water.
2. Place the cap securely on each film can.
3. Stack 0, 1, 2, 3, or 4 water-filled film cans on top of each of the seed-filled cans (Figure 1.1).
4. Check every hour to see if seed swelling has forced the cap off any of the seed-filled cans.
5. Graph the time it takes to force the cap off each seed-filled can against the number of water-filled cans stacked on the seeds.
6. If you have a balance, weigh a water-filled can. Determine how much mass is pressing on each of the seed-filled cans. Graph the time it takes to force the cap off each seed-filled can against the total mass of the water-filled cans stacked on each seed-filled can.
7. As a variation, use different salt concentrations in each seed-filled can and leave off the stacks of water-filled cans. Graph the time it takes to force the cap off each seed-filled can against the salt concentration.

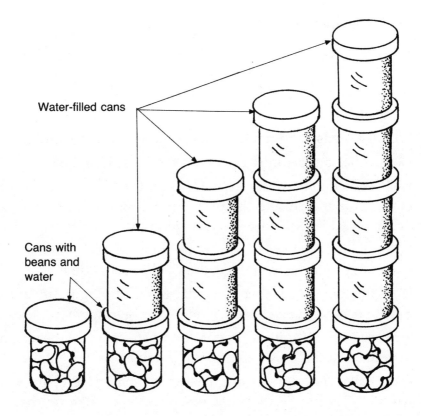

Water-filled cans

Cans with
beans and
water

Figure 1.1. Setup to measure the force produced by imbibing bean seeds confined in plastic film cans.

Expected Results

As the imbibing seeds swell, they will force the cap off the film can. The more water-filled cans stacked on the cap, the longer it should take to force the cap off.

Why?

If dry seeds or wood are confined and then wettened, large pressures can develop. The pressure is caused by the adhesion between water molecules and the dry seed. The dry seed pulls water molecules into the seed, and the seed volume increases. An ancient technique to cut stone blocks uses dry wooden pegs driven tightly into holes drilled in the rock. The wood pegs are then wettened, and the rock splits due to the expansion of the peg as it imbibes water. Some people believe that the stone blocks for the pyramids were cut using this method.

Further Investigations

1.　How do different types of seeds compare in the preceding experiments? Try types of bean seeds that differ in size. Try a gymnosperm seed (pine) versus a monocot seed (corn) versus a dicot seed (bean).

2.　Will dry seeds still imbibe if they are ground in a blender?

3.　Does water temperature affect the rate of seed imbibition?

4.　Will seeds imbibe water vapor? Try putting 20 g of dry seeds in a plastic cup and sealing the plastic cup inside a jar holding a 1-cm depth of water. Measure the change in mass of the seeds.

5.　Will fully imbibed seeds, which are then air-dried, imbibe as much water and at the same rate as they did the first time?

6.　Does imbibing seed, then immediately air-drying it, affect the germination percentage? (See Chapter 2 on germination percentage.)

7.　Will seeds imbibed in salt water germinate as well as seeds imbibed in plain water? (See Chapter 2 on germination percentage.)

8.　Does salt concentration affect seed imbibition? Prepare solutions with 10, 20, and 40 ml of table salt (sodium chloride) per liter of water. Do the time-course experiment for each of the three salt concentrations and for plain water as a zero salt control treatment. Graph the final seed volume or seed mass against the salt concentration.

FURTHER READING

Raven, P. H., R. F. Evert, and S. E. Eichhorn. 1992. *Biology of plants*. 5th ed. New York: Worth Publishers.

CHAPTER **2**

Germination Percentages of Different Types of Seeds

When sowing seeds, it is important to know how many of the seeds will germinate. This information is the **germination percentage,** which is the number of seeds in every 100 that will sprout when given proper conditions of water, oxygen, temperature, and light. Seed packets usually have a germination percentage printed on them.

Seeds can be germinated on moist filter paper in a Petri dish (Figure 2.1). If you do not have Petri dishes, you can use either a plastic plate enclosed in a plastic bag or the bottom half of a 2-liter soda bottle (see Appendix 1), a paper towel, and clear plastic wrap (Figure 2.2). The dry paper towel is placed on the inside wall of the soda bottle and wetted. It sticks to the wall due to adhesion, as do small seeds that are placed on the wet paper towel.

In this project, you will learn how to measure the germination percentage for several packets of seeds. You will also determine whether or not the germination percentage on the packet agrees with your measurements.

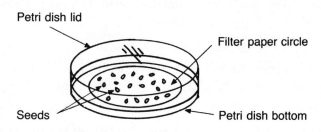

Figure 2.1. Method of germinating seeds on moist filter paper in a Petri dish.

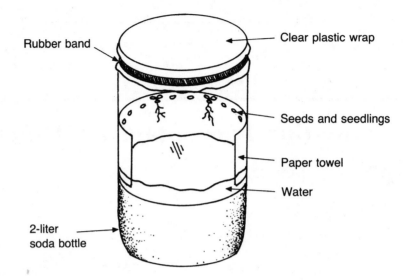

Rubber band

Clear plastic wrap

Seeds and seedlings

Paper towel

Water

2-liter
soda bottle

Figure 2.2. Method of germinating seeds on paper towels in the bottom of a 2-liter plastic soda bottle.

Purpose

To determine the germination percentage of different types of seeds.

Materials

☐ plastic Petri dishes with lids (enough to hold 20 of each type of seed)
☐ filter paper circles to fit the Petri dishes
☐ water
☐ packages of several types of flower or vegetable seeds
☐ forceps (optional)

Procedure

1. Place a filter paper circle in the bottom of each Petri dish, and saturate the filter paper with water.
2. For each type of seed, place 20 seeds on the moist filter paper in a Petri dish (Figure 2.1). Use a forceps for small seeds. For large seeds, you may need two or three Petri dishes for each batch of 20 seeds. Close each Petri dish after all the seeds have been added.
3. Place the Petri dishes in a warm location if possible.
4. Every day, check the filter paper to be sure it is still saturated.

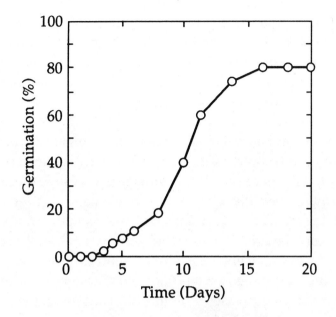

Figure 2.3. General change in seed germination percentage over time.

5. Every day, count the number of seeds germinated. Consider a seed to have germinated when the radicle has emerged 1 cm from the seed. Multiply the number of seeds germinated by 5 to obtain the germination percentage. You multiply by 5 because you started with 20 seeds. If all 20 seeds germinate, the germination percentage is 100, which equals 5 × 20.

6. Stop the experiment when the germination percentage does not change for 4 days in a row.

7. Graph the germination percentage against the time in days (Figure 2.3).

Expected Results

The germination percentage will often be lower than on the seed package and will usually differ among different kinds of seeds.

Why?

Most of our main food and forest crops are grown from seed, including wheat, rice, corn, oats, barley, oak, and pine. Because seeds are so important in food and forest production, there are laws that regulate seed quality. Most states have seed testing laboratories that test samples of all seeds sold in that state. The seed testing laboratory measures the **seed purity** by determining the percentage by mass of **pure seed,** other seed, and nonseed materials in the seed sample. The pure seed is the type of seed listed on the seed label. A batch of seeds cannot be sold if it contains too few pure seeds or too many seeds of certain weeds. The seed purity is listed on the seed label.

The seed testing laboratory also measures the germination percentage. The date of the test and the germination percentage are both placed on the seed label. If the germination percentage is very low, the seed cannot be sold. Farmers need the germination percentage so they know how many seeds to sow. A farmer must sow twice as many seeds if the germination percentage is 50% than if the germination percentage is 100%.

The germination percentage is also a useful test for home gardeners who collect their own seed or who use old packages of seed. The germination percentage decreases with time depending on the type of seed and how it was stored. The best way to store most types of seed is to keep the dry seeds in an airtight container at 0°C. Some types of seed will not germinate after even one year of storage, but most common flower, vegetable, and grain seeds will still germinate after three or more years of proper storage. However, the germination percentage will often be lower than that listed on the seed label.

One reason why your germination percentage may not have been as high as on the seed package is that seed testing laboratories have special chambers for germinating seeds. The chambers keep the temperature constant. The temperature in your germination test was probably not constant.

Look on the seed packet at the date of the germination test. How long has it been since the germination test? It was probably several months to a couple of years. The germination percentage decreases with time. This decrease is greater if seeds are not stored dry and at low temperatures. Seed packets in the store or home are usually stored at room temperature. Seeds packets are often not waterproof; therefore, the seeds may absorb water from the air by imbibition (see Chapter 1). Usually, the higher the water content of a seed, the less time it will survive in storage.

Further Investigations

1. How does the seed storage temperature affect the germination percentage? Try storing batches of 20 seeds in airtight plastic con-

tainers. Store them at room temperature (about 21°C), in the refrigerator (about 4°C), and in the freezer (about −18°C). Determine the germination percentage after storage for a month or longer.

2. Does using a desiccant (drying agent) during seed storage affect the germination percentage (see the Further Reading section)?

3. How does the temperature during germination affect the germination percentage? Try placing the Petri dishes for the germination percentage test at room, refrigerator, and freezer temperatures.

4. Does fertilizer increase the germination percentage? Try adding a fertilizer solution instead of water to half of the Petri dishes.

Further Reading

Bugbee, B. 1989. Storing seeds: Use desiccants to keep them dry. *Fine Gardening* 5: 57–59.

Duncalf, W. G. 1976. *The Guinness book of plant facts and feats.* Enfield, Middlesex, U.K.: Guinness Superlatives.

CHAPTER 3

Cotyledon Removal and Seedling Growth

In dicot seeds, the two cotyledons contain energy for the embryo. This energy can be in the form of starch, protein, or oil. The cotyledons are carried above the ground by the elongating hypocotyl when a typical dicot seed germinates. However, sometimes the cotyledons are torn off when the hypocotyl moves through the soil. In some types of dicots, the cotyledons remain below ground and only the epicotyl elongates during germination. In this project, you will use dicot seeds that carry their cotyledons aboveground.

In this project, you will examine how the loss of one or both cotyledons affects the seedling. You will also determine if one cotyledon per seed is better than two half cotyledons per seed.

Purpose

To determine how seedlings are affected by removal of the cotyledons.

Materials

☐ dicot seeds, such as bean, chia, radish, or Wisconsin fast plant
☐ water
☐ plastic pots
☐ potting soil
☐ saucers
☐ single-edge razor blade
☐ marking pen
☐ light source

☐ ruler

☐ balance

Procedure

1. Cover the seeds with water and soak overnight.

2. Sow two or three seeds per pot of moist potting soil. A variety of plastic containers can be used, such as yogurt cups, Styrofoam cups, and Styrofoam egg cartons. Be sure each container has a drainage hole in the bottom. Cover the seed to a depth about equal to twice its width and water gently.

3. Place the pots on saucers in a warm (about 20–25°C), well-lighted place.

4. When the cotyledons are above the soil, thin to one seedling per pot by using the razor blade to cut off the weaker seedlings at the soil level.

5. Remove both cotyledons from one-third of the plants using a razor blade. Label those pots 0.

6. Remove one cotyledon from another one-third of the plants. Label those pots 1.

7. Do not remove anything from the last third of the plants. Label those pots 2.

8. Measure the height of each plant daily.

9. One to two weeks after removing the cotyledons, cut off each plant at the soil level.

10. Measure the fresh mass of each plant with a balance.

11. Graph the plant height against the time in days. Graph the plant fresh mass versus the number of cotyledons (0, 1, and 2).

Expected Results

Removal of the cotyledons usually reduces the growth of the seedling.

Why?

The cotyledons are filled with compounds that provide energy and building materials for the young seedling. Removing the cotyledons slows the growth of the seedling because it has to rely on **photo-**

synthesis alone for energy and building materials. Photosynthesis is the process whereby plants produce **carbohydrates** using light energy, water, and carbon dioxide. The plant can use the carbohydrates as a source of energy or can convert them into any other compounds it needs for growth. Much of the structural part of the plant is made of **cellulose,** a carbohydrate that people cannot digest.

Further Investigations

1. Does removing half of both cotyledons have the same effect as removing one whole cotyledon?

2. Can the addition of fertilizer to the plants with one or two cotyledons substitute for the removal of one or both cotyledons?

3. Does cotyledon removal have as much effect on plant growth in small-seeded plants, such as chia, as it does in large-seeded species, such as bean?

4. The cotyledons often make up most of the seed in many dicots, such as bean. Do larger seeds, which have larger cotyledons, grow faster than smaller seeds of the same type?

5. Can you provide a substitute for a cotyledon? Try cutting off a cotyledon and then dip the cut end into a sugar solution (5 ml/liter of water).

CHAPTER 4

Determining the Best Sowing Depth

If a seed is not sown deeply enough, the roots may be too close to the soil surface. This might cause the seedling to fall over. Seeds not sown deep enough may dry out and are more likely to be eaten by birds. A seed may not have enough energy to push through the soil and reach the light if it is sown too deeply. What is the best sowing depth for a seed?

In this project, you will determine the best depth for seed sowing. You will discover if this best sowing depth is related to the size or the type of the seed. You can also examine if the type of potting soil affects the best depth for sowing.

Purpose

To determine the best sowing depth for different types of seeds.

Materials

☐ seeds of various sizes, such as bean, corn, chia, and Wisconsin fast plant

☐ water

☐ clear, 2-liter plastic soda bottles cut 20 cm from the bottom, with a hole in the bottom for drainage (see Appendix 1)

☐ potting soil

☐ ruler

☐ aluminum foil

Procedures

For large seeds

1. Soak large seeds, such as corn and bean, overnight in water to speed germination.

2. Fill the 20-cm soda bottle one-fourth full with moist potting soil.

3. Place two seeds next to each other on top of the potting soil and pressed against the clear wall of the soda bottle. Orient both seeds in the same direction. For example, place corn seeds with the pointed end down, and place bean seeds with the hilum facing up.

4. Press both seeds gently into the potting soil until the top of the seed is level with the soil surface.

5. Add a 2-cm layer of potting soil, and firm gently by tapping the soda bottle on the tabletop.

6. Repeat steps 3 to 5 until the soil surface is 3 cm from the top of the soda bottle. Do not sow seeds directly above one other. The seeds should form a stairstep pattern (Figure 4.1). The top of the last two seeds should be even with the soil surface.

7. Measure the depth of each pair of seeds from the soil surface to the top of the seed. Sowing depths should be about 0, 2, 4, 6, 8, 10, and 12 cm.

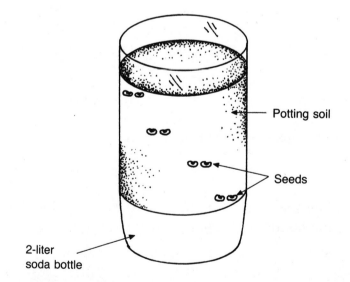

Figure 4.1. Setup to determine the best sowing depth by placing seeds in a stairstep pattern.

For small seeds

1. Fill the 20-cm soda bottle one-half full with potting soil.

2. Follow steps 3 to 6 above, except use 0.5-cm layers of potting soil in step 5 above.

3. Measure the depth of each pair of seeds from the soil surface to the top of the seed. Sowing depths should range from 0.5 to 7 cm.

For all seeds

1. After sowing, cover the sides of the soda bottles with aluminum foil to exclude light.

2. Every day, remove the aluminum foil and measure the length of each seedling's root and shoot. Note how many days after sowing it takes each seedling to break through the soil surface. This is the day of emergence.

3. Graph the day of emergence against the depth of sowing.

4. Graph the seedling shoot length against the depth of sowing.

Expected Results

Seeds planted too shallowly or too deeply will not grow as well as seeds planted at an intermediate depth. The best sowing depth for small seeds should be less than the best sowing depth for big seeds.

Why?

If the seed is sown too deeply, the seedling will use too much energy before it gets above the soil surface. The seedling will die if it uses all its energy before it reaches the light. A seed sown too deeply may not get enough oxygen or may not have sufficient strength to push through the soil. The depth of sowing is also important for seeds that require light for germination and for seeds in which light prevents germination.

Further Investigations

1. Is there an ideal orientation of the seed during sowing? A bean seed can be sown at least five different ways: (1) hilum up, (2) hilum down, (3) seed on its side with the hilum to the side, (4) seed vertical with the hilum above the micropyle, and (5) seed vertical with the hilum below the micropyle.

2. Does the type of potting soil affect the ideal depth of sowing? Try using several different materials, such as sand, perlite, vermiculite, and sphagnum peat moss.

3. What happens if there is a barrier above or below the seed? Try putting small stones just above or just below a seed sown in the soda bottle. Watch what the seedling root or shoot does when it runs into the stone.

CHAPTER 5

Carbon Dioxide Production and Oxygen Consumption by Germinating Seeds

Most seeds get their energy from starch, protein, or oils stored in the cotyledons or endosperm. The starch, protein, and oil in seeds is important food for animals. High-starch seeds include wheat, corn, and rice. High-protein seeds include beans, chia, and nuts. High-oil seeds include sunflower, safflower, and chia.

In this project, you will see whether or not germinating seeds change the content of carbon dioxide and oxygen in the air. You will use a calcium hydroxide solution to test for carbon dioxide. A burning stick will serve as a test for oxygen.

Purpose

To determine how oxygen and carbon dioxide levels change when seeds respire.

Materials

- ☐ calcium hydroxide (available from a garden center or biological supply company)
- ☐ teaspoon
- ☐ 5 clear glass soda bottles with caps
- ☐ water
- ☐ 90 navy bean seeds
- ☐ marking pen
- ☐ stove

☐   cookpot

☐   coffee filter

☐   plastic funnel

☐   50-ml glass beakers, or small, clear glass bottles

☐   plastic drinking straw

☐   2-hole, #3 rubber stoppers

☐   scissors

☐   aquarium tubing

☐   5- or 10-ml plastic syringes without needles

☐   wooden splints or wood popsicle sticks

☐   matches

Procedure

1. Add about 5 ml (1 teaspoonful) of calcium hydroxide to a soda bottle filled with water, cap the bottle tightly, and shake it. *WARNING: Calcium hydroxide is slightly caustic, so handle it carefully and do not get it in your eyes, mouth, or nose.*

2. Shake the calcium hydroxide solution periodically to help it dissolve.

3. Soak two batches of 30 navy bean seeds in water for 24 hours.

4. Place one batch of the soaked seeds in a separate soda bottle, and cap it tightly. Label this bottle "soaked."

5. Boil the other batch of soaked seeds for 10 minutes.

6. After the water has cooled, remove the boiled seeds.

7. Place the batch of boiled seeds in a separate soda bottle, and label it "boiled."

8. Place a batch of 30 dry bean seeds in a separate soda bottle, and cap it tightly. Label this bottle "dry."

9. Tightly cap one empty soda bottle, and label it "control."

10. Let the bottles remain capped for 12–24 hours.

11. Test each bottle for the presence of carbon dioxide as follows:

    a. Pour the calcium hydroxide solution through the coffee filter held by the funnel over a glass beaker or bottle. The top 10 cm of a 2-liter plastic soda bottle can be used as a funnel (Figure 5.1).

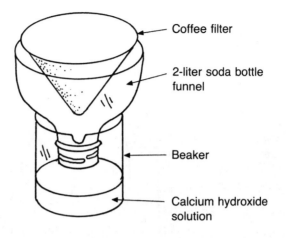

Coffee filter

2-liter soda bottle funnel

Beaker

Calcium hydroxide solution

Figure 5.1. Setup to filter calcium hydroxide solution using a funnel made from the top of a 2-liter plastic soda bottle.

b. Test some of the filtered calcium hydroxide solution from the beaker or bottle by breathing into it through the drinking straw. It should turn milky white because of the carbon dioxide in your breath.

c. Replace the soda bottle cap on each bottle, one at a time, with a two-hole, #3 rubber stopper. Put a 30-cm piece of aquarium tubing in the top of one hole and a 5-cm piece of aquarium tubing in the top of the second hole. Stick the tip of a 5- or 10-ml plastic syringe into the 5-cm piece of tubing (Figure 5.2). There should also be a 15-cm piece of aquarium tubing in the bottom of the stopper hole holding the 5-cm tubing. Separate pieces of tubing are stuck in the top and bottom of the same hole because it is difficult to pull a single piece of tubing all the way through the hole.

d. Place the end of the 30-cm tubing in the beaker of calcium hydroxide solution.

e. Pour 60 ml of water into the plastic syringe in order to force the air out of the bottle and bubble it through the calcium hydroxide solution. Use the same amount of water for each bottle so that equal amounts of gas are bubbled through the calcium hydroxide solution.

f. Rate the amount of milkiness in each calcium hydroxide solution as follows: no milkiness (−), slight milkiness (+), moderate

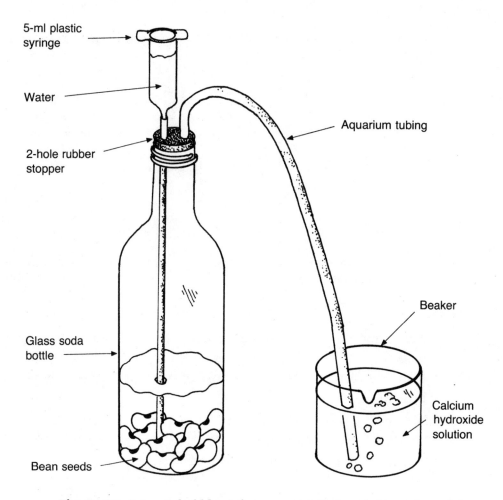

Figure 5.2. Setup to bubble air from germinating seeds through a calcium hydroxide solution to test for carbon dioxide.

milkiness (++), or high milkiness (+++). Compare the other samples to the breath test results (see step 11b), which should be rated high milkiness.

12. With an adult helper, test each bottle for the presence of oxygen as follows:

   a. Place each bottle in a sink and uncap it.

   b. Lower a burning wood splint or wood popsicle stick into the bottle.

   c. Rate the test as positive (+) for oxygen if the wood continues to burn when lowered into the bottle. Rate the test as negative (−) if the flame goes out when the wood is lowered into the bottle.

A negative test does not necessarily mean that there is no oxygen, only that the amount of oxygen is much less than normal.

Expected Results

Soaked seeds should give a negative oxygen test and a positive carbon dioxide test. The no seed (control), boiled seed, and dry seed treatments should give a positive oxygen test and negative carbon dioxide test.

Why?

Germinating seeds cannot use photosynthesis, so they get their energy from the starch, protein, or oils stored in the seed. **Respiration** is the process that produces energy when starch, protein, or oils are broken down into carbon dioxide and water. Respiration requires oxygen and produces carbon dioxide. Our bodies, like plants, respire all the time. The respiring seeds use the oxygen in the bottle and convert it to carbon dioxide.

Dry seeds respire at a very low rate compared to germinating seeds because they are not growing. The low rate of respiration of dry seeds enables them to conserve their stored energy. Boiled seeds do not respire because they have been killed. Respiration only occurs in living seeds. After a few days, boiled seeds may give a positive carbon dioxide test due to microbes growing on the boiled seeds.

Bubbling carbon dioxide into a calcium hydroxide solution turns it milky white due to the formation of solid calcium carbonate. The milky appearance will disappear as the solid calcium carbonate settles. Calcium hydroxide is also called hydrated lime or slaked lime, which is sold in garden centers. Calcium carbonate is found in blackboard chalk, eggshells, and seashells.

The burning of wood requires oxygen. Like respiration, the burning of wood uses oxygen, releases energy, and produces carbon dioxide and water. The flaming stick continues to burn when it is lowered into a bottle filled with air. The flame goes out when the burning stick is lowered into the bottle containing the wet seeds. There is much less oxygen in the bottle because the seeds have converted it to carbon dioxide.

Further Investigations

1. Will soaked seeds that are left under water give a positive carbon dioxide test and negative oxygen test?

2. Will dry and soaked bean seeds die if they are placed in a freezer? If so, how long does it take to kill them? Try using the calcium hydroxide test to see whether or not the seeds survive freezing.

3. How will soaked seeds kept in a capped bottle compare to soaked seeds planted in potting soil after one or two weeks? Try to explain any differences in growth.

4. Do germinating seeds produce heat? Try putting a handful of soaked bean seeds in a small thermos bottle. Measure the air temperature with a thermometer. Stick the thermometer into the handful of seeds, and see if the temperature increases. Use dry or boiled seeds as a control.

PART II

Plants and Water

Water is the main component of most plants. Nonwoody plants, such as lettuce, may be about 95% water by mass. Plants use water in several ways. Plants do not have bones to support them so they use water to help keep them rigid or turgid. A tire will go flat if it does not have enough air in it. Plants use water instead of air, so a plant will "go flat," or wilt, if it does not have enough water in it. The pressure inside plant cells that keeps them turgid is **turgor pressure.**

Plants use water in many chemical reactions, such as photosynthesis. Most chemicals in the plant are dissolved in water, and water movement transports other chemicals from place to place in the plant. To help cool the plant, water evaporates from it, a process called **transpiration.**

Transpiration occurs when water evaporates through tiny holes in the leaf called **stomates,** or **stomata.** Each stomate, or stoma, is surrounded by two **guard cells.** The stomate opens when the guard cells gain water and swell; the stomate closes when the guard cells lose water and shrink. The stomates open in the light, which allows carbon dioxide to get into the leaf for photosynthesis; the stomates close in the dark in most plants because photosynthesis does not occur in the dark. Stomates also close in the light if the roots cannot absorb enough water. Net photosynthesis stops when stomates close in the light.

Plants absorb most water through their roots. The water moves up the stems and into the leaves in tiny tubes called **xylem,** which is a **vascular tissue** because it functions in moving substances through the plant. About 90% of the water absorbed by a plant is transpired. Thus, plants need much more water, relative to their size, than animals.

FURTHER READING

Kramer, P. J. 1983. *Water relations of plants.* San Diego, Calif.: Academic Press.

CHAPTER 6

Stomatal Density of Different Leaves

Stomates, or stomata, are small openings in the leaf surface. Each stomate is surrounded by two specialized cells, called guard cells (Figure 6.1). A stomate opens when the guard cells absorb water and swell. The stomate closes when the guard cells lose water and shrink. Stomates are very small. An open stomate on a tomato leaf has an area of about 0.000078 mm².

In this project, you will use a microscope to examine stomates. A difficulty in seeing stomates through the microscope is that leaves are many cell layers thick. You can peel off the outer layer of leaf cells and look at them under the microscope, but it is difficult to peel off the outer cell layer in most types of leaves. Instead, for this experiment you will make an impression of the leaf surface and its stomates. The impression is made by painting the leaf with clear fingernail polish. When the polish has dried, you can peel it off and view it under the microscope.

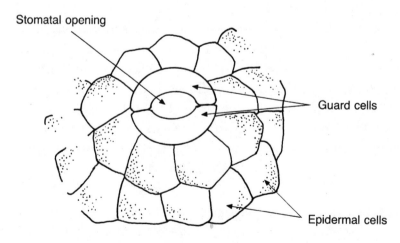

Figure 6.1. Structure of a stomate.

You can then determine the number of stomates per unit area of the leaf, which is the **stomatal density.**

Purpose

To determine the number and location of stomates on several types of leaves.

Materials

- ☐ plants with hairless leaves, such as Wandering Jew, coleus, and Wisconsin fast plant
- ☐ clear fingernail polish
- ☐ transparent tape
- ☐ microscope slides
- ☐ microscope
- ☐ flat, clear plastic millimeter ruler (optional)

Procedure

1. Paint about a 1-by-1-cm area on the bottom of a hairless leaf with clear fingernail polish. Avoid breathing the fingernail polish fumes. Leaves with a lot of hairs do not work well. Leaves can remain attached to the plant when the polish is applied. Dried pressed leaves can also be used.
2. Allow the polish to dry in a well-ventilated room.
3. Place a piece of transparent tape over the dry fingernail polish and press firmly.
4. Slowly peel the tape from the leaf. The dry fingernail polish should come up with the tape.
5. Stick the tape to a clean, dry microscope slide. Press it firmly in place.
6. Examine the slide under a microscope. Use high power to see the stomates.
7. Make sure the leaf impression fills the entire field of view of the microscope. Determine if the stomates are open or closed. Note the shape and size of the stomate and its guard cells.
8. Count the number of stomates visible in the entire field of view.

Determine the stomatal density by dividing your count of stomates by the area of the microscope's field of view. Ask your teacher or parent for this information or calculate it as follows:

a. Set the microscope at about 100× magnification.

b. Place the plastic millimeter ruler on the microscope stage and bring it into focus.

c. Move the ruler so the inner edge of a millimeter line is just visible on the left side of the microscope's field of view. The edge of the ruler should be exactly in the middle of the field of view. In other words, the edge of the ruler divides the field of view in half.

d. Measure the diameter of the field of view in millimeters. Estimate the fraction of a millimeter between the last line on the right and the edge of the field of view—for example, 1.4 mm.

e. Divide the diameter by 2 to get the radius. Calculate the area of the field of view with the equation for the area of a circle: area = Pi × radius squared, where Pi = 3.14. For example, if the diameter = 1.4 mm, then the radius = 1.4 ÷ 2 = 0.7 mm; thus, area = 3.14 × 0.7 × 0.7 = 1.54 mm².

9. Repeat steps 1 to 8 with several types of leaves. Compare stomate size, stomate shape, and stomatal density for these various leaves. Be sure the microscope magnification is set the same for each leaf; otherwise, the comparisons are not valid.

10. Repeat steps 1 to 8 with the top surfaces of different types of leaves. Compare the stomates from the tops to those from the bottoms.

Expected Results

Most dicot leaves have stomates mainly on the lower surface. The stomatal density is very high, up to 100,000 stomates/cm².

Why?

A single leaf may have many millions of stomates. Scientists have found as many as 100,000 stomates/cm² of leaf surface. Most dicot leaves have more stomates on the underside of the leaf than on the top side. Many types of leaves have no stomates at all on the top side of the leaf. One reason for fewer stomates on the top of many types of leaves is the waxy covering, or **cuticle,** on the top of the leaf. The cuticle helps the leaf shed rain and prevents excessive evaporation. Floating leaves, such as water lily, only have stomates on the top surface of the leaf. Upright leaves, such as corn and grasses, have about the same number of stomates on both sides of the leaves.

Further Investigations

1. How do monocot and dicot stomates compare in terms of size, shape, stomatal density, and distribution on the top and bottom of the leaf? How do angiosperm and gymnosperm stomates compare? Do primitive plants, such as ferns, have stomates? Try the fingernail polish method to answer these questions.

2. Do flower petals or stems have stomates? Try the fingernail polish method on petals and young stems.

3. What do **hydathodes** look like? Hydathodes are modified stomates on the tips or edges of leaves. Water droplets often form at the tips or edges of leaves during warm, humid nights when the soil is moist. These droplets are not dew but come from inside the plant. The formation of these droplets is called **guttation.** Guttation occurs through the hydathodes. Try the fingernail polish method on the edges of several leaves, and see if you can find some hydathodes.

4. Are stomates open or closed in the dark and in the light? Try putting fingernail polish on leaves at night or leaves kept in the dark.

Further Reading

Esau, K. 1977. *Anatomy of seed plants.* 2d ed. New York: John Wiley and Sons.
Meyer, B. S., D. B. Anderson, R. H. Bohning, and D. G. Fratianne. 1973. *Introduction to plant physiology.* New York: D. Van Nostrand.

CHAPTER 7

Environmental Factors Affecting Transpiration

Transpiration is the loss of water from the plant. Most transpiration occurs through stomates (see Chapter 6). About 90% of the water absorbed by roots will be transpired. Thus, plants require huge amounts of water. Transpiration is most easily measured in potted plants by their loss of mass. The volume of transpiration is easily calculated from the loss in mass because 1 g of water has a volume of 1 ml.

In this project, you will measure the amount of transpiration of a potted plant. You will also determine how wind, humidity, and closing the stomates affect transpiration.

Purpose

To determine the rate of transpiration in potted plants exposed to different conditions.

Materials

- [ ] 5 equal-sized, single-stem plants (such as geranium or coleus) in pots
- [ ] marking pen
- [ ] water
- [ ] plastic bags that will fit over the pots
- [ ] twist ties
- [ ] light source
- [ ] petroleum jelly
- [ ] small electric fan
- [ ] balance

Procedure

1. Label the potted plants A, B, C, D, and E.

2. Water each plant until water drains out the hole in the bottom of the pot.

3. After drainage stops, place the plant pots in a plastic bag. Close the bag securely with a twist tie around the base of the stem (Figure 7.1). Be careful not to damage the plant stem.

4. Place plants A to D where they receive about the same temperature and the same amount of light. Use a fluorescent light system if available (see Appendix 2).

5. Designate plant A as the control.

6. Coat the bottoms of the leaves of plant B with a thin layer of petroleum jelly.

7. Coat the tops of the leaves of plant C with a thin layer of petroleum jelly.

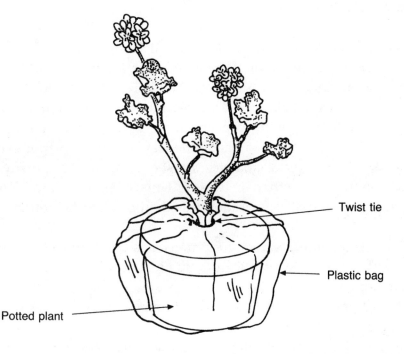

Figure 7.1. Method of enclosing a pot in a plastic bag to prevent evaporation from the potting soil.

8. Focus the electric fan so that it blows directly on plant D. Be sure it does not blow on the other plants.

9. Place plant E in the dark in a location with about the same temperature as the other plants.

10. Weigh each plant container, and record the mass in grams. Write down the time of weighing.

11. Every hour, reweigh each pot. Continue weighings for at least 8 hours.

12. Calculate the transpiration in grams of water per hour by subtracting each mass from the one preceding it. For example, if pot mass is 250 g at hour 0 and 241 g at hour 1, the transpiration rate is 250 − 241 = 9 g/hour.

13. Graph the transpiration rates in grams per hour against the time in hours.

Expected Results

The transpiration rates, from highest to lowest, should be D, A, C, B, and E.

Why?

The **relative humidity** inside a normal leaf is very close to 100%. The relative humidity is the ratio of the amount of water vapor in the air to the amount of water vapor the air can hold. The relative humidity outside a leaf is usually below 100%. Molecules move from areas of high concentration to areas of low concentration. Thus, water molecules will usually move from inside to outside the leaf. The water vapor exits the leaves mainly through the stomates. Most dicot leaves have many more stomates on the bottom than on the top. Thus, coating the top of the leaf with vaseline reduces transpiration less than coating the bottom of the leaf.

If there is no wind, water vapor molecules build up just outside the stomate. This raises the relative humidity just outside the leaf and slows transpiration. Wind blows away these water molecules and increases the rate of transpiration. In the dark, stomates of most types of plants close, so transpiration is low.

Scientists often say that transpiration is a wasteful process that occurs because stomates must let carbon dioxide into the leaf for photosynthesis. However, transpiration may benefit the plant because evaporation of water cools the leaf. Also, the rapid movement of water through the xylem carries mineral nutrients from the roots to the leaves.

Further Investigations

1. Do **antitranspirants** improve plant growth? Antitranspirants—for example, Wilt-Pruf®—are sprayed on plants to reduce transpiration when plants are transplanted or during winter. Water in the soil is often frozen in winter, so evergreen leaves can die if they transpire too much. Try spraying an antitranspirant on a plant and then measuring the rate of transpiration and the rate of growth compared to unsprayed plants.

2. What is the effect of salt concentration in the soil on transpiration? Try watering a potted plant with a solution of table salt (5 ml salt/liter). Measure the transpiration rate of the plants watered with salt water and plain water.

3. What is the **transpiration ratio** of a potted plant? The transpiration ratio is the number of grams of water transpired per gram of plant dry mass produced. To measure the transpiration ratio, plant a small seed in a pot of potting soil. After the seed germinates, cover the soil surface with plastic as in step 3 of the "Procedure" section. Measure the milliliters of water added to the soil over several weeks. Do not water the plant so much that water drains from the pot. Cut off the plant at the soil surface, and dry the plant in the sun or in a 70°C oven. Measure the mass of the dried plant in grams. Divide the milliliters of water added to the plant by the plant dry mass in grams to get the transpiration ratio. For example, if you added 1500 ml of water to the plant and the dry mass of the plant is 3 g, the transpiration ratio is $1500 \div 3 = 500$. The transpiration ratio is expressed as a single number—for example, 500, rather than 500:1—because the second number is always 1.

CHAPTER 8

Leaching from Potted Plants

Leaching is the loss of water and dissolved substances from the soil **root zone.** The root zone is the depth of soil in which the roots grow. The root zone in a pot usually is the whole pot. The solution that leaches is called the **leachate.** Dissolved substances in the leachate include fertilizers, salts, and pesticides. Leaching is a problem when fertilizers and pesticides get into well water that is used for drinking. Another problem with leaching is that it wastes water and fertilizer because they are leached below the roots. One benefit of leaching is the removal of excess salts that can harm plants.

In this project, you will study leaching from potted plants. You will examine how leaching can benefit the plant and how leaching can harm the environment.

Purpose

To determine how the amount of leaching affects the growth of potted plants.

Materials

- ☐ six 2-liter plastic soda bottles (see Appendix 1)
- ☐ cork borer
- ☐ 4 bases from 2-liter soda bottles (see Appendix 1)
- ☐ 4 equal-sized potted plants
- ☐ marking pen
- ☐ soluble houseplant fertilizer, such as Miracle-Gro®
- ☐ graduated cylinder
- ☐ water

☐ table salt
☐ ruler
☐ balance (optional)
☐ electrical conductivity meter (optional)
☐ 35-mm plastic film can (optional)
☐ distilled water (optional)
☐ plastic cups (optional)

Procedure

1. Cut four 2-liter plastic soda bottles in half 15 cm from the bottom (see Appendix 1).

2. Cut a hole in the center of the four soda bottle bases with the cork borer, and place each base on top of one of the four half-bottles from step 1 (Figure 8.1).

3. Set each of the four potted plants in one of the bases from step 2.

4. Label the plants A, B, C, and D.

5. In a 2-liter bottle, prepare a fertilizer solution by following the directions on the fertilizer package. Label this solution AB. Irrigate

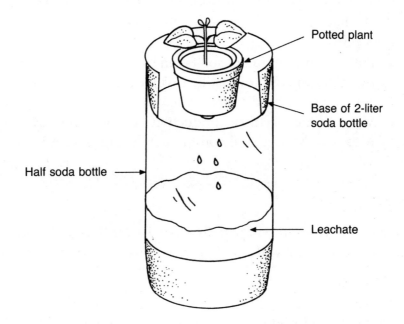

Potted plant

Base of 2-liter soda bottle

Half soda bottle

Leachate

Figure 8.1. Setup to catch leachate from a potted plant in the bottom half of a 2-liter plastic soda bottle.

plants A and B with solution AB using a graduated cylinder to keep track of the volume. For plant A, apply enough solution so that the volume applied is about twice the volume that leaches. For plant B, apply only enough solution so that no leaching occurs. If a little leaching does occur, apply the leachate to the soil during the next irrigation.

6. In another 2-liter bottle, prepare a fertilizer solution by following the directions on the fertilizer package and adding 5 ml of table salt. Label this solution CD. Irrigate plants C and D with solution CD using a graduated cylinder to keep track of the volume. For plant C, apply enough solution so that the volume applied is about twice the volume that leaches. For plant D, apply only enough solution so that no leaching occurs. If a little leaching does occur, apply the leachate to the soil during the next irrigation.

7. For all plants, keep track of the volume of solution added to each pot and the volume of solution that leaches from each pot. The solution that leaches will be caught in the soda bottle. Empty the soda bottle into the graduated cylinder to measure the leachate.

8. Grow the plants for at least 4 weeks, and irrigate the pots at least twice a week.

9. Measure plant height. Measure shoot fresh mass if you have a balance.

10. If you have an electrical conductivity meter, do the following for each pot:
    a. Put one film can full of soil taken from the middle of the pot and two film cans full of distilled water into a plastic cup.
    b. Mix the soil and distilled water and allow the soil to settle.
    c. Measure the electrical conductivity of the water layer.
    d. Measure the electrical conductivity of solution AB and solution CD.

Expected Results

Plants that are leached will usually grow better than unleached plants. Plants that receive salt will not grow as well as plants that do not get salt.

Why?

Plant growth is reduced if there is too much salt in the soil. This occurs because the salt competes with the roots for water. When the soil water is salty, the roots cannot absorb it as easily. Thus, the plant suffers from a lack of water when there is a lot of salt in the soil solution. Soluble fertilizers are salts, too, so too much fertilizer in the soil can have an effect similar to table salt.

When the soil is not leached, the salt can build up in the soil solution. Leaching will flush the excess salt out of the soil, which can benefit the plant. In field crops, leaching of fertilizer into well water can make the water unsafe to drink. Therefore, it is better to avoid leaching by not adding excess fertilizer and using water that is low in salt.

Electrical conductivity is an easy way to measure the salt level in the soil. The higher the electrical conductivity, the higher the soil salt level.

Further Investigations

1. What is the ideal **leaching fraction** for growing potted plants? The leaching fraction is the volume of solution leached divided by the volume of solution applied to the plant. This project used leaching fractions of 0.5 and 0. Try growing plants with leaching fractions of 0, 0.2, 0.4, and 0.6 to determine which leaching fraction produces the best plant growth. The ideal leaching fraction will depend on several factors including the type of plant and the environmental conditions.

2. How much of the salt you applied was leached? Save the leachate from plant C. Evaporate the solution and determine how many milliliters of salt remain. Boiling the leachate in an old cookpot or in a microwave oven will speed the rate of evaporation. Another method to speed evaporation is to increase the area of solution exposed to the air. This can be done by pouring the leachate into pans lined with plastic wrap. When the salts have dried, they can be easily removed from the pan by removing the plastic wrap. Compare the milliliters of fertilizer and salt applied to the milliliters of salt leached.

Further Reading

Hershey, D. R., and S. Sand. 1993. Electrical conductivity. *Science Activities* 30(1): 32–35.

Ku, C. S. M., and D. R. Hershey. 1991. Leachate electrical conductivity and growth of potted poinsettia with leaching fractions of 0 to 0.4. *Journal of the American Society for Horticultural Science* 116: 802–6.

CHAPTER 9

Acid Rain Injury to Plants

A major environmental problem in many parts of the world, especially in the eastern United States, is **acid rain.** Rain is considered acid rain when the pH is below 5.6. Acid rain is caused by the burning of fossil fuels, such as oil and coal. Burning releases sulfur dioxide and nitrogen oxide gases, which react with ozone in the atmosphere to form sulfuric acid and nitric acid. These acids are washed out of the atmosphere by rain and other forms of precipitation. Many forests and lakes have been badly damaged by acid rain.

In this project, you will determine the effect of artificial acid rain on plant growth. The term **acid precipitation** is sometimes used in place of acid rain. Precipitation includes all forms of water that fall from the sky including snow, sleet, hail, and fog.

Purpose

To determine the effect of acid precipitation on plant growth.

Materials

☐ seeds, such as bean
☐ 2 plastic pots
☐ potting soil
☐ light source (see Appendix 2)
☐ marking pen
☐ plastic wrap
☐ distilled water
☐ 2-liter plastic soda bottle (see Appendix 1)
☐ medicine dropper
☐ nitric or sulfuric acid

☐ pH meter or pH paper (available from Carolina Biological Supply [see Appendix 5])

☐ marking pen

☐ 2 spray bottles

Procedure

1. Sow the seeds in pots with moist potting soil, water them, and place them in bright light.

2. When the bean seedlings have fully expanded their first pair of leaves, label one plant container "acid" and the other "control."

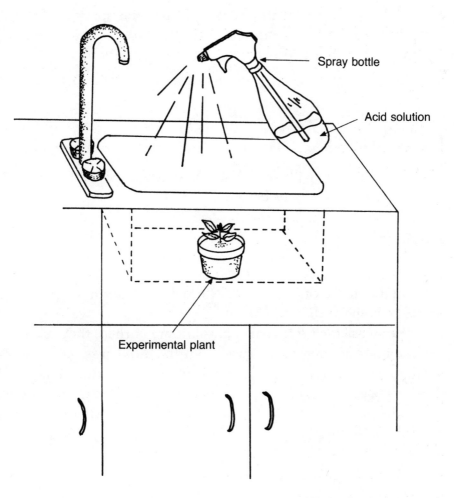

Spray bottle

Acid solution

Experimental plant

Figure 9.1. Method of spraying a potted plant with an acid solution to simulate acid rain.

3. Cover each half of the soil surface in each pot with a separate piece of plastic wrap. The stem should poke through between the two pieces of plastic.

4. Pour 1 liter of distilled water into a clean 2-liter soda bottle.

5. With adult help, add a drop of nitric acid or sulfuric acid to the distilled water. Mix by swirling the water in the bottle.

6. Test the water pH using a pH meter or pH paper. If the pH is above 3, add more acid. If it is below 3, add more distilled water. Test the water pH until it is about 3. Then pour it into a spray bottle.

7. Place the control plant in a sink, and use a second spray bottle containing distilled water to mist the leaves. Let the leaves dry, and then return the control plant to its growing location.

8. Place the acid plant in a sink, and use the spray bottle with the pH 3 solution to mist the leaves (Figure 9.1). Let the leaves dry, and then return the acid plant to its growing location.

9. Observe any differences in growth and leaf color between the two treatments.

Expected Results

The plant sprayed with the pH 3 solution will be badly damaged. Its leaves will turn yellow or brown.

Why?

Rain will normally be slightly acid because carbon dioxide gas will dissolve in it to make carbonic acid. Normal rainwater has a pH of 5.6 due to the carbonic acid.

When fossil fuels like coal or gasoline are burned, they release sulfur dioxide and nitrogen oxide gases into the air. These gases react with sunlight, ozone, and water vapor to form nitric and sulfuric acids. When these acids collect in the rain, the pH of the rain is much less than 5.6. When the pH is below 5.6, it is called acid rain or acid precipitation. The low pH can harm plants.

Further Investigations

1. How low does the pH of rain have to be to cause damage to leaves? Try spraying plants using distilled water with pHs of 3, 3.5, 4, and 4.5.

2. Can plants be sprayed with something to protect them from acid rain? Try spraying the top of the leaves with an antitranspirant, like Wilt-Pruf.

3. What effect does acid rain have on soil pH? Try irrigating potted plants with acid solutions and see if it affects plant growth (see Chapter 16).

4. Do you have acid precipitation in your area? Try setting out plastic containers to collect precipitation. Measure the pH of the water you collect.

5. Is nitric acid more harmful than sulfuric acid? Try spraying plants with nitric acid solution, sulfuric acid solution, or a mixture of both acids. Be sure the pH of each solution is the same, and be sure to use distilled water as a control.

Further Reading

Mohnen, V. A. 1988. The challenge of acid rain. *Scientific American* 259(2): 30–38.

PART III

Light and Photosynthesis

Light is extremely important to plants in many ways. Most plants are **autotrophic,** meaning that they obtain their energy from photosynthesis, which is the process whereby plants use light energy to produce sugar from carbon dioxide and water. In contrast, animals are **heterotrophic,** meaning they get energy by consuming other organisms rather than using light directly as their energy source. Plants will die if they do not get enough light because they cannot photosynthesize enough. However, too much light will injure or kill some types of plants. Light may promote or inhibit seed germination. Plant parts sense light and may respond by growing toward or away from it (phototropism). The color or quality of light can affect many aspects of plant growth. For example, plants receiving far-red light grow much taller than plants receiving blue light. The length of the daily light period also influences many aspects of plant growth including the formation of plantlets and flowering.

Photosynthesis is the most important plant process that requires light. Photosynthesis requires water and carbon dioxide, a colorless, odorless gas that occupies 0.035% of the air by volume. Photosynthesis also produces oxygen as a byproduct. We require oxygen to breath, so we must thank photosynthetic plants for providing it. In return, we breath out the carbon dioxide that plants require for photosynthesis. Before photosynthesis began on earth about 3 billion years ago, the air contained no oxygen. Today, the air has 21% oxygen, by volume.

Chlorophyll-containing cells are also required for photosynthesis. **Chlorophyll** is the pigment responsible for the green color of most leaves. It absorbs the light energy required for photosynthesis. Chlorophyll is located in **chloroplasts,** which are tiny packets within plant cells. Photosynthesis occurs in the chloroplasts.

Photosynthesis is a complex process but is usually summarized by the following equations:

$$\text{carbon dioxide} + \text{water} \xrightarrow[\substack{\text{chlorophyll-containing} \\ \text{cells}}]{\substack{\text{light and mineral} \\ \text{nutrients}}} \text{glucose} + \text{oxygen} + \text{water.}$$

In words, the equation is carbon dioxide and water are converted into glucose and oxygen when chlorophyll-containing cells receive light and mineral nutrients. The water and mineral nutrients enter through the plant roots. The carbon dioxide enters through the leaves.

It is important to remember that photosynthesis requires not just chlorophyll but also chlorophyll-containing cells. Chlorophyll alone is not enough. There are dozens of other compounds required in addition to chlorophyll. The equation is a bit misleading because large amounts of **glucose** are not found in photosynthesizing leaves. Glucose is a sugar with six carbon atoms. Instead, most plants join glucose molecules end to end in long chains or **polymers.** These long chains of glucose molecules are called **starch,** which is easily detected in most leaves (see Chapter 11). Glucose molecules joined in a different way form cellulose, which is the major component of plant cell walls.

Scientists like to use chemical symbols instead of words. Using chemical symbols, the preceding equation for photosynthesis becomes

$$6CO_2 + 12H_2O^* \xrightarrow[\substack{\text{chlorophyll-containing} \\ \text{cells}}]{\substack{\text{light and mineral} \\ \text{nutrients}}} C_6H_{12}O_6 + 6O_2^* + 6H_2O.$$

This equation shows the numbers of each molecule required per glucose molecule produced. Notice that 12 water molecules (H_2O) are required (left side of equation) and 6 are produced (right side of equation). The reason 12 H_2O are needed is that each is split during photosynthesis. The 12 oxygen atoms (O) combine to form the 6 oxygen molecules (O_2). The asterisk following the "O" in H_2O^* is to distinguish it from the "O" in CO_2. The asterisk appears with the O_2^* on the right side of the equation, indicating that the O_2^* comes from the H_2O^* on the left side. Thus, the oxygen (O) in the glucose ($C_6H_{12}O_6$) comes from the carbon dioxide (CO_2). The carbon (C) in the glucose also comes from CO_2. Only the hydrogen (H) in glucose comes from the water (H_2O). Both carbon and oxygen atoms have a much greater mass than a hydrogen atom does. This means that 93% of the mass of glucose comes from the carbon dioxide. Plants gain mass largely on a diet of air containing a small concentration of carbon dioxide.

FURTHER READING

Asimov, I. 1989. *How did we find out about photosynthesis?* New York: Walker.

Govindgee and W. J. Coleman. 1990. How plants make oxygen. *Scientific American* 262(2): 50–55, 58.

Salisbury, F. B., and C. W. Ross. 1991. *Plant physiology.* 4th ed. Belmont, Calif.: Wadsworth.

Taiz, L., and E. Zeiger. 1991. *Plant physiology.* New York: Benjamin/Cummings.

Wilkins, M. 1988. *Plantwatching: How plants remember, tell time, form relationships and more.* New York: Facts on File Publications.

CHAPTER 10

Photosynthesis in Leaf Disks

In this project, you will use leaf disks cut with a plastic drinking straw to examine the process of photosynthesis. The leaf disks are placed in a syringe of baking soda solution. Baking soda is sodium bicarbonate, which provides a source of carbon dioxide.

When the leaf disks are exposed to light, oxygen is produced via photosynthesis. The oxygen bubbles formed within the leaf cause it to float to the top. Placing the floating leaf disks in the dark stops photosynthesis, and the oxygen in the disks is consumed via respiration. The leaf disks then sink to the bottom.

Purpose

To determine how environmental conditions affect the rate of photosynthesis in leaf disks.

Materials

- ☐ black 35-mm film cans
- ☐ baking soda
- ☐ water
- ☐ plastic drinking straw
- ☐ seedling of Wisconsin fast plant
- ☐ 5-ml plastic syringe without needle
- ☐ fluorescent light source, such as a desk lamp
- ☐ stopwatch, or watch with second hand

Procedure

1. Barely cover the bottom of a 35-mm film can with baking soda. Fill the film can with water and replace the cap. Shake the film can to dissolve the baking soda.

2. Use the plastic straw to cut four leaf disks from one seedling. All four leaf disks should remain inside the plastic straw stacked one on top of each other.

3. Remove the cap and plunger from the syringe. Gently blow the leaf disks out of the straw into the bottom of the syringe barrel (Figure 10.1).

4. Replace the plunger in the barrel of the syringe. Slowly push in the plunger most of the way. Leave about 1 ml of air in the syringe so the leaf disks are not crushed.

5. Place the tip of the syringe in the baking soda solution. Pull out the plunger and draw 4 ml of solution into the syringe.

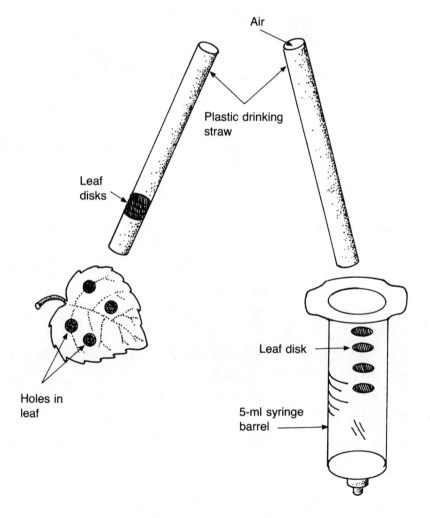

Figure 10.1. Cutting leaf disks with a plastic drinking straw and transferring them to a syringe.

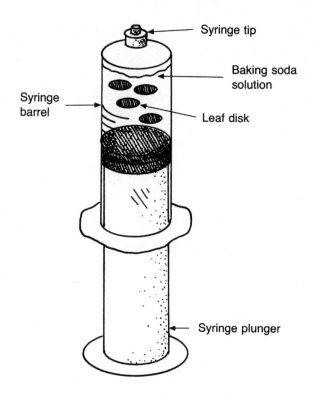

Figure 10.2. System to observe photosynthesis of leaf disks in a 5-ml plastic syringe.

6. Turn the syringe with the tip up and push the plunger in to expel any air still in the syringe (Figure 10.2).

7. Hold your finger over the opening in the tip of the syringe while you pull down on the plunger. This creates a vacuum in the solution, which pulls the air out of the leaf disks. The leaf disks should all sink. Tap the side of the syringe until they do sink.

8. Place the syringe, tip up, under a fluorescent light. The leaf disks should be about 5 cm from the light bulb.

9. Record the time that passes before each leaf disk floats to the top of the syringe.

10. When all disks have floated, darken the syringe by placing it in a dark room or box. Another method is to invert a black 35-mm film can over the tip of the syringe.

11. Record the time it takes for each leaf disk to sink.

12. After all the disks have sunk, place the syringe so the disks are 10 or 20 cm from the fluorescent bulb. Repeat steps 9 to 11.

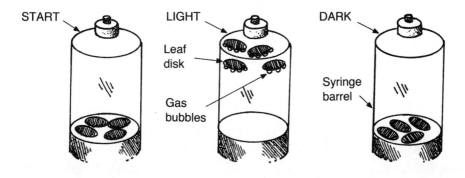

Figure 10.3. Leaf disks float in the light due to oxygen production during photosynthesis and sink in the dark when oxygen is consumed during respiration.

Expected Results

Leaf disks rise when they produce oxygen via photosynthesis (Figure 10.3). In the dark, the leaf disks sink when the oxygen is consumed in respiration. The rate of photosynthesis decreases as the light level decreases (distance between the disks and the light source increases).

Why?

The rate of photosynthesis can be determined most readily by measuring the plant's consumption of carbon dioxide or the plant's production of oxygen. This project measures oxygen production indirectly by determining how long it takes the leaf disk to float. The faster a disk floats, the more oxygen it must have produced.

Oxygen production is easy to observe with aquatic plants. A classic classroom experiment uses elodea (*Elodea canadensis*), a common aquarium plant. A bunch of cut elodea stems are placed in a beaker of water under a submerged funnel. A water-filled test tube is then inverted over the neck of the funnel. The rate of photosynthesis is estimated by counting the number of air bubbles that are released from the cut end of the elodea stem and travel up into the test tube. However, it is rather tedious to count the air bubbles. Also, the method may be inaccurate because the bubbles are not all the same size.

Further Investigations

1. Will nongreen areas of **variegated** (multicolored) leaves function in this leaf disk photosynthesis system? Try cutting disks from the green and nongreen areas of a variegated leaf, such as coleus.

2. Will the leaf disk photosynthesis system work without baking soda in the solution?

3. What colors of light are most effective for photosynthesis? Try covering the syringe with various colors of cellophane.

4. What other types of leaves will work in the leaf disk photosynthesis system? Try several other types of leaves, such as gymnosperm, dicot, and monocot leaves. You cannot cut disks from most gymnosperms because they have needle leaves. Try ginkgo (*Ginkgo biloba*) leaves; ginkgo is a gymnosperm with flat leaves.

5. How does temperature affect the rate of photosynthesis? After step 8, try heating the syringe in warm water or cooling it by placing it in a refrigerator. Keep the syringe tip up in a beaker to prevent water from leaking out.

6. How does mineral nutrient deficiency affect the rate of photosynthesis? Try using leaf disks cut from nutrient-deficient plants (see Chapter 20).

Starch Testing of Leaves

Starch is a familiar substance found in bread, pasta, rice, potato, and many other common foods. Starch is made of thousands of glucose molecules joined together in long chains. During photosynthesis, starch forms in most types of leaves.

In this project, you will test leaves for starch with iodine, which is available at drugstores. The iodine test is an easy way to test leaves for starch. A pale yellow iodine solution reacts with starch to form a black or blue-black color. To test a leaf for starch, the leaf is first boiled in water to kill the cells and swell the starch. Next, the leaf is boiled in denatured alcohol. The hot alcohol dissolves the chlorophyll, leaving the leaf white. The bleached leaf is then treated with iodine to test for starch. The leaf turns blue-black if starch is present.

Purpose

To test for starch in leaves and to determine the conditions required for starch to form in leaves.

Materials

☐ plant with thin, green leaves, such as bean, geranium, fuchsia, or Wisconsin fast plant

☐ black construction paper or aluminum foil

☐ paper clip

☐ light source

☐ stove

☐ cookpot

☐ water

- ☐ 2 large Pyrex® test tubes
- ☐ denatured alcohol
- ☐ metal food can
- ☐ metal test tube holder
- ☐ pot holder
- ☐ tincture of iodine
- ☐ medicine dropper
- ☐ rice, bread, or corn starch
- ☐ plastic Petri dish
- ☐ white paper

Procedure

1. Place a green-leaved plant in complete darkness overnight. Test a leaf from the plant for starch (steps 4a to 4h). If the test is positive, keep the plant in the dark for a longer time.

2. Place a penny-size circle of black construction paper or aluminum foil on both sides of a green leaf. Secure them with a paper clip (Figure 11.1).

3. Place the plant in bright light for 3–4 hours.

4. Test both the covered and an uncovered leaf for starch as follows:

   a. Place each leaf in boiling water for several minutes.

   b. Take each leaf out of the water, roll it, and place it in a test tube.

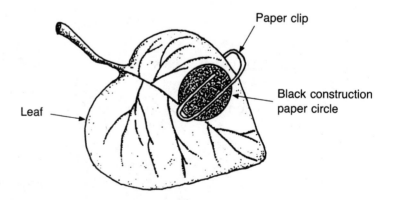

Figure 11.1. Method of attaching a construction paper circle to a leaf with a paper clip.

c. Cover each leaf in its test tube with denatured alcohol.

d. Place a clean metal food can in the cookpot. The metal food can is used so the test tube will sit upright in the boiling water (Figure 11.2). Fill the food can and pot with water. Bring the water to a boil on a stove.

e. When the water is boiling, turn off the stove. *WARNING: Alcohol is flammable so keep it away from the open flame of a gas stove.*

f. With adult help, hold the test tube with the metal test tube holder and the pot holder. Lower the test tube into the boiling water in the metal food can. The test tube should stand upright when it leans against the side of the can (Figure 11.2). The alcohol should turn green as the chlorophyll dissolves.

g. With adult help, remove the test tube from the boiling water and allow it to cool to room temperature. Pour off the alcohol. If the leaf is still green, add more denatured alcohol and repeat steps 8d to 8g.

h. Fill the test tube with water for a few minutes. Soaking the leaf in water removes the brittleness caused by the alcohol.

i. Dilute the tincture of iodine 20-fold by adding 20 drops of water to each drop of tincture of iodine. Place a drop of the diluted iodine on a grain of rice, a bread crumb, or pinch of corn starch to test it. The rice, bread, or corn starch should turn black.

j. Lay the bleached leaf in the Petri dish. Place the Petri dish over a white piece of paper so a positive starch test is easily visible.

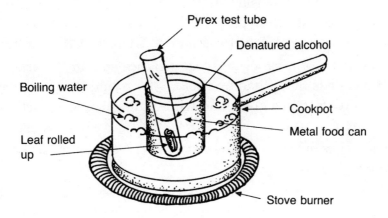

Figure 11.2. Double-boiler method of removing chlorophyll from a leaf by boiling in denatured alcohol.

k. Cover the surface of the bleached leaf with the iodine solution. A positive starch test is a black or blue-black color. The starch test is either positive (+) or negative (−).

5. If desired, preserve the leaves in jars of 50% alcohol. The blue-black color will soon fade; however, the color can be restored by a new application of iodine solution.

Expected Results

The uncovered leaf will turn completely black, but the leaf with black construction paper over it will have a white circle on it where the construction paper was.

Why?

Leaf starch is usually depleted at night because photosynthesis does not occur in the dark. Thus, testing leaves for starch is a way to determine whether or not photosynthesis is occurring or has recently occurred. To be sure leaves were depleted of starch at the beginning of this experiment, the plant was kept in the dark overnight until it gave a negative starch test. The leaf covered with the circle of black construction paper did not receive light in that area, so it could not photosynthesize and produce starch. That is why there was a white circle on the covered leaf after the starch test.

Further Investigations

1. How does petroleum jelly affect starch formation? Try covering the top or bottom of a starch-depleted leaf with petroleum jelly before putting the plant in bright light for 3 hours. Then, test the leaf for starch.

2. Do all parts of a variegated leaf produce starch? A variegated leaf normally has more than one color—for example, green with white edges. Try testing a green and white variegated leaf, such as coleus, for starch.

3. Do all types of leaves produce starch? Try the starch test on leaves of barley (*Hordeum vulgare*).

4. Which types of seeds contain starch? Try soaking several types of seeds in water overnight to soften them. With adult help, carefully cut a soaked seed in half. Place a drop of iodine solution on the cut surface to test for starch.

CHAPTER 12

Plant Carbon Dioxide Deficiency

Plants require carbon dioxide for photosynthesis. In fact, most of the dry mass of the plant comes from carbon dioxide (see the beginning of Part III). Most living organisms also give off carbon dioxide via respiration when they produce energy. A common misconception is that plants only respire at night. The truth is that plants respire all the time. They only seem not to respire during the day because they use more carbon dioxide during photosynthesis than they produce during respiration.

In this project, you will grow one plant with and one plant without carbon dioxide. Each plant will be grown in a system built from six 2-liter plastic soda bottles. The plants are sealed inside one of the bottles. The system to deprive plants of carbon dioxide uses two bottles of calcium hydroxide solution (see Chapter 5). The control system uses plain water instead of calcium hydroxide solution.

Purpose

To determine how the lack of carbon dioxide affects plant growth.

Materials

- [ ] twelve 2-liter plastic soda bottles (see Appendix 1)
- [ ] sewing needle
- [ ] ruler
- [ ] washed gravel
- [ ] 2-hole, #3 rubber stoppers
- [ ] scissors
- [ ] aquarium tubing
- [ ] water
- [ ] calcium hydroxide

☐ graduated cylinder
☐ 2 aquarium valves
☐ 2 aquarium pumps
☐ cotton or polyester floss
☐ 2 plants in small pots
☐ 2 twist ties
☐ plastic electrical tape
☐ balance

Procedure

1. Prepare twelve 2-liter soda bottles as described in "Preparing Soda Bottles," in Appendix 1.

2. Mark two of the 2-liter bottles at 18 cm from the bottom, but only draw the mark three-fourths of the way around the bottle (see "Cutting Soda Bottles," in Appendix 1).

3. With adult help, cut the two bottles at the mark with the razor blade and scissors (see Appendix 1). The two parts of the bottles will remain attached by the uncut area, which serves as a hinge (Figure 12.1A).

4. With adult help, use the sewing needle to punch one hole in the upper and lower parts of the two bottles (Figure 12.1A). Each hole should be 0.5 cm from the cut edge.

5. Fill the two cut bottles about one-third full of washed gravel to provide drainage for the plants.

6. Divide the 12 bottles into two systems of 6 bottles, with the two cut bottles positioned as the fifth bottle in each system. Insert the aquarium tubing into the two-hole rubber stoppers (see Figure 12.1B). Insert the stoppers in the soda bottles.

7. Slip a plastic drinking straw over the tip of the aquarium tubing that is inside all but the first of the six bottles in each system.

8. Fill the second, third and fourth bottles in each system 90% full of water. Fill the sixth bottle in each system 20% full of water.

9. Add 5 ml of calcium hydroxide to the second and third bottles in one system. This is the system that will cause carbon dioxide deficiency.

10. Attach the first bottle in each system to an aquarium valve and

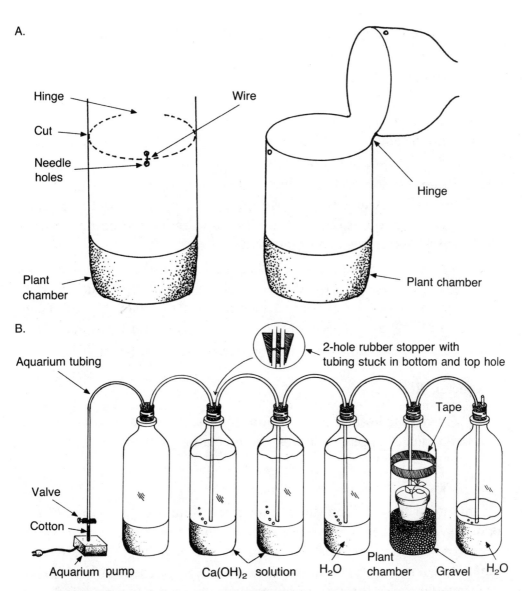

Figure 12.1. A. Method of preparing the plant chamber bottles. B. System for depriving plants of carbon dioxide built with 2-liter plastic soda bottles.

aquarium pump. Be sure to put a small piece of cotton or polyester fiber floss in the air tubing attached to the aquarium pump. This serves to filter dirt from the air so the aquarium valve is not clogged.

11. Place a small potted plant in each of the cut bottles. Strip the covering off the wires of the twist ties. Align the two small holes in each cut bottle and tie them shut with the wire from the twist tie. Make sure the cut edges fit firmly together. Cut off the excess wire at the knot.

12.   Seal the cut edges of each bottle with electrical tape, taping over the wire closure. The seal must be airtight.

13.   Arrange each system of six bottles so that the two bottles with a plant are side by side in a well-lighted spot.

14.   Plug in the aquarium pump and adjust the airflow for each system with the aquarium valve. There should be a gentle bubbling in each bottle, including the last one.

15.   Water will accumulate in the bottle with the plant due to condensation. If the water rises above the gravel, open the bottle top and siphon off the excess water. Close and reseal the bottle after the excess water has been removed.

16.   Observe the plants daily. Plants deprived of carbon dioxide will not grow, and some of their leaves will die. After carbon dioxide deficiency symptoms become apparent, measure plant height. Then, cut off the plants at soil level. Measure the plant fresh mass.

17.   Test leaves from each plant for starch (see Chapter 11).

Expected Results

The plant in the system with calcium hydroxide will not grow and will probably die. This plant's leaves will give a negative starch test.

Why?

Plants must have carbon dioxide for photosynthesis to occur. Without photosynthesis, plants cannot make the compounds they use for energy and for building materials. Photosynthesis naturally generates oxygen that is used in respiration to create carbon dioxide. Thus, carbon dioxide continually cycles through photosynthesis to respiration and back.

Carbon dioxide is also produced when we burn fuels, such as oil, natural gas, coal, and wood. In the last few decades, the production of carbon dioxide on earth has exceeded the carbon dioxide used in photosynthesis. This has caused a slow increase in the carbon dioxide concentration in the air. Some scientists are predicting that the increased carbon dioxide will cause **global warming,** a rise in the earth's temperature. Higher temperatures might cause more polar ice to melt and lead to flooding of seacoasts. One possible solution to global warming is to plant more trees. The trees will absorb more carbon dioxide from the air and store it as wood, which is composed largely of cellulose.

An increase in the carbon dioxide concentration may increase the rate of photosynthesis. From a plant perspective, carbon dioxide is very valuable and oxygen is mainly a waste product. This is just the opposite of the human viewpoint. The current high level of oxygen in the air (21% by volume) has built up over hundreds of millions of years due to photosynthesis. There was no oxygen gas in the air before photosynthesis began.

While we are worrying about too much carbon dioxide in the air, greenhouse growers must also worry about too little carbon dioxide in their greenhouses. During cold weather, greenhouses are kept closed so that plants do not freeze. However, plants can quickly use up all the carbon dioxide in a closed greenhouse. Therefore, greenhouse growers often add carbon dioxide to their greenhouses.

Further Investigations

1. If two plant chambers are placed one after the other in the same system, will the plant in the second chamber grow better or worse than the first plant?

2. Will plant growth increase if the carbon dioxide is increased above the normal level in air? Try putting soaked seeds or decaying plant matter in the second and third bottles, in place of water.

3. Do plants require carbon dioxide at night? Try changing the calcium hydroxide bottles as follows. Just before each daily dark period, replace the two calcium hydroxide bottles with water-filled bottles. Just before each daily light period, replace the water-filled bottles with the calcium hydroxide bottles. This deprives the plant of carbon dioxide during the light but not during the dark.

4. Can you use this system to demonstrate that plants respire in the light as well as in the dark? Try letting the air exiting the plant chamber bubble through a filtered solution of calcium hydroxide (see Chapter 5).

5. How can plants in a sealed terrarium survive for years when no carbon dioxide is being added?

Further Reading

Bazzaz, F. A., and E. D. Fajer. 1992. Plant life in a CO_2-rich world. *Scientific American* 266(1): 68–74.

Hershey, D. R. 1992. Plants can't do without CO_2. *Science Teacher* 59(3): 41–43.

CHAPTER 13

C-3, C-4, and CAM Photosynthesis

There are three major types of photosynthesis. **C-3 photosynthesis** is so named because the first compound produced has three carbons. Similarly, **C-4 photosynthesis** produces a four-carbon compound first. CAM stands for **crassulacean acid metabolism.** It is named for the Crassulaceae, a family of succulent plants. Plants with CAM photosynthesis produce an organic acid first.

In this project, you will investigate C-3 and C-4 photosynthesis and determine how well C-3 and C-4 plants compete with each other for carbon dioxide. The plants need to be grown in a mineral soil, such as coarse sand or perlite, and watered with a nutrient solution. Regular potting soils release carbon dioxide and may interfere with the experiment.

Purpose

To determine how plants with different systems of photosynthesis compete with each other.

Materials

- [ ] four 2-liter plastic soda bottles with 1 cap (see Appendix 1)
- [ ] perlite
- [ ] 10 corn seeds
- [ ] 10 oat seeds
- [ ] water
- [ ] nutrient solution (see Appendix 3)
- [ ] fluorescent light system (optional [see Appendix 2])
- [ ] plastic electrical tape

Procedure

1. Cut off two soda bottles just above the base and save the top parts.

2. Cut off two soda bottles 23 cm from the bottom and save the bottom parts (see Appendix 1).

3. Fill the 23-cm soda bottle bottoms one-third full of perlite.

4. Plant five oat and five corn seeds in each of the two soda bottle bottoms.

5. Irrigate the perlite with a nutrient solution. Be careful not to overwater because there is no drainage hole in the bottom.

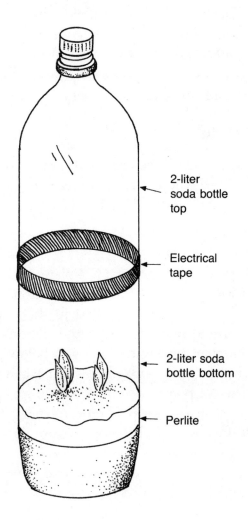

Figure 13.1. Two-liter plastic soda bottle chamber for comparing growth of C-3 and C-4 plants.

6. Place the containers under a light bank or in bright light.

7. When the seedlings are 5 cm tall, pinch off all but one oat and one corn seedling per soda bottle. Try to keep the best seedling of each type.

8. Place each soda bottle top on each soda bottle bottom (Figure 13.1). Tape the two soda bottle pieces together with electrical tape to form a chamber. Be sure the plant chamber seal is airtight.

9. Tightly seal one plant chamber with the soda bottle cap. Leave the other plant chamber uncapped.

10. Place the containers in bright light, either sunlight or fluorescent. Do not place them in direct sunlight or they may overheat.

11. Water the plants in the uncapped chamber if required. The plants in the capped chamber should not need water.

12. After two weeks, compare the growth of both types of plants in the capped and uncapped chambers.

Expected Results

In the sealed chamber, the oat seedling will die, and the corn seedling will survive. However, it will not grow much larger. Both seedlings will grow larger in the unsealed chamber.

Why?

Oat is a C-3 plant, and corn is a C-4 plant. C-4 plants can photosynthesize with a lower carbon dioxide concentration than C-3 plants can. When the corn and oat are sealed in the plant chamber, they both photosynthesize at first. However, the carbon dioxide concentration in the plant chamber quickly decreases to below the level needed by the oat plant. The oat plant stops photosynthesizing and growing but keeps respiring. The oat plant's respiration produces carbon dioxide.

The corn plant can keep photosynthesizing because it can use the low carbon dioxide concentration. It can also use the carbon dioxide produced by respiration of the oat plant. The oat plant will keep respiring until it runs out of stored energy and dies. The corn plant will stay alive but may not grow much because there is not much carbon dioxide available.

Another type of photosynthesis is CAM, or crassulacean acid metabolism. CAM photosynthesis is found in many cactus and other plants that live in dry areas. CAM plants open their stomates in the dark and absorb carbon dioxide from the air. The carbon dioxide is made into organic acids using stored energy. In the light, the stomates close. The plant uses light and the stored carbon dioxide for photosynthesis. CAM plants do not grow very fast but they use very little water. Common CAM plants are the pineapple and devil's backbone.

Further Investigations

1. How do C-3 and CAM plants compete with each other? Try this project using a C-3 plant like oat and a CAM plant like devil's backbone.

2. How does CAM photosynthesis affect the transpiration rate? Try removing two leaves from a devil's backbone plant. Weigh each leaf, and then place one leaf in the dark and place one leaf in continuous light. Reweigh the leaves every few days. Which leaf loses the most water?

3. How do CAM and C-4 plants compete with each other for water? Try sowing a corn seed and planting a devil's backbone plantlet in the same pot of potting soil. Keep it well watered until the corn has sprouted and is several centimeters tall. Then, do not water the pot, and see which plant survives the longest.

4. How does CAM photosynthesis work? Try the experiments suggested in the article listed in the Further Reading section.

Further Reading

Friend, D. J. C. 1990. Plant eco-physiology: Experiments on crassulacean acid metabolism, using minimal equipment. *American Biology Teacher* 52: 358–63.

PART **IV**

Soils and Fertilizers

Soil can be defined simply as a medium for the growth of plant roots. Soil has four main functions for plant growth. They can be remembered by associating the first letter of each one with a letter in the word soil. The four functions are support, oxygen, ions, and liquid. Soil supports, or anchors, the plant. It also provides oxygen; ions, or mineral nutrients; and liquid, or water, for the roots.

Soils consist of three parts: solids, water, and air. A typical field soil will have about 50% solids, by volume. The other 50% would be holes or pores that would be filled with air and water. The pores are like a complex maze of tunnels. They allow gases to enter and exit the soil. Most pores are filled with air in a dry soil. Most pores are filled with water in a flooded soil. About half the pores hold water and half hold air in a moist soil, so a well-watered soil in the field is about 50% solids, 25% air, and 25% water, by volume.

Potting soils are made for plants growing in containers or pots. They are different from field soils because they often have only 15% solids, by volume, and 85% pores. Potting soils hold up to 70% water and 15% air, by volume, shortly after they have been watered. The large percentage of water is needed because of the small volume of potting soil in a pot. A plant in the field has a much larger soil volume than one in a pot.

Potting soils are more fragile than field soils and must be handled gently. Rough handling can greatly reduce the size of the pores so the potting soil holds less air and water. Potting soils should not be pressed down too hard when planting. Tapping the pot on a tabletop is okay for firming the soil in a pot. Watering the soil also helps to settle the soil. Many packaged potting soils are air dry when opened. Hot tap water should be added to moisten them before you plant. Hot water wets the soil more easily than cold water. The potting soil should not be so wet that water runs out when a handful of soil is squeezed. Some potting soils are very dusty when dry. Avoid breathing the dust, as it can be irritating. Keep the dust down by misting the potting soil with water from a spray bottle until you can moisten it.

Soil is important for plant growth but it is not essential. Plants grow fine without soil in hydroponics (see Part V). However, most land plants do grow in soil. A common idea is that plant roots eat soil. Unlike animals and people, most plants do not really eat anything. Plant roots absorb water and mineral nutrients from the soil. They use these substances in photosynthesis to create all the molecules they need for energy, growth, and reproduction. Plants are food for people, but most plants do not use food themselves. Fertilizers are often called "plant food," but it is incorrect to call fertilizers food.

A fertilizer is defined as a material that contains one or more of the mineral nutrients required by plants. Most types of plants require 14 mineral nutrients. The **macronutrients** are required in fairly large amounts. The six mineral macronutrients are nitrogen, phosphorus, potassium, calcium, magnesium, and sulfur. Plants also require carbon, hydrogen, and oxygen, which are considered macronutrients, too. They are not mineral nutrients because they come from carbon dioxide and water. The eight **micronutrients** are needed in fairly small amounts. The micronutrients are iron, boron, manganese, copper, zinc, molybdenum, chlorine, and nickel. Some types of plants also need silicon, sodium, or cobalt.

Fertilizers, like seeds (see Chapter 2), are regulated by law. A fertilizer label must have a **guaranteed analysis,** which consists of three numbers—for example, 5-10-10. The numbers are the percentage by mass of nitrogen, phosphorus pentoxide, and potassium oxide in the fertilizer. Most fertilizers are inorganic, meaning they were manufactured or mined. Common inorganic fertilizers are superphosphate, potassium nitrate, and ammonium nitrate. Organic fertilizers come from living organisms. Organic fertilizers include bone meal, blood meal, and fish emulsion.

Organic farmers often claim that plants grown with organic fertilizers are better than plants grown with inorganic fertilizers. There is no scientific support for this claim. The mineral nutrients in both types of fertilizers are exactly the same. Organic fertilizers are better for the environment because they are recycled, they do not require fossil fuels to manufacture, and they are not mined.

FURTHER READING

Brady, N. C. 1984. *The nature and properties of soils*. 9th ed. New York: Macmillan.

Marschner, H. 1986. *Mineral nutrition of higher plants*. London: Academic Press.

California Fertilizer Association. 1990. *Western fertilizer handbook: Horticulture edition*. Danville, Ill.: Interstate Publishers.

Plant Growth in Various Potting Soils

Egyptian hieroglyphics show plants being grown in pots more than 3000 years ago. Today, plants are often grown in containers or pots, as in many of the projects in this book. Is it really necessary to use a packaged potting soil to grow plants in pots? Why not just use soil from the garden? In this project, you will answer these questions by growing potted plants in both garden soil and potting soil.

Purpose

To determine the effect of different types of potting soils on plant growth.

Materials

- ☐ 2 plastic pots
- ☐ potting soil
- ☐ garden soil
- ☐ marking pen
- ☐ seeds
- ☐ saucers
- ☐ water
- ☐ light source
- ☐ ruler

Procedure

1. Fill a plastic pot with moist potting soil and another with moist garden soil.

2. Label each pot with the type of soil.

3. Sow seeds in each pot, and set each pot in a saucer.

4. Water the pots, and set them in bright light in a warm location.

5. Remove all but one seedling from each pot after they have their first true leaves.

6. Observe and measure plant height daily for several weeks.

Expected Results

Plants will often grow better in packaged potting soil than in garden soil.

Why?

A typical well-watered soil in a garden has about 50% solids, 25% air, and 25% water, by volume. If this soil is put in a small pot, it will hold much more water. After watering, a garden soil in a pot would still have about 50% solids, by volume, but it would have closer to 50% water and 0% air. Roots need air so plants often do not grow well in pots of garden soil.

Another reason that plants do not grow well in pots of garden soil is that garden soil often carries plant diseases or plant pests, such as insects. Packaged potting soils usually do not have plant diseases or insects. Packaged potting soils also have been supplied with a small amount of fertilizer and have a proper pH. Garden soils may or may not have proper amounts of fertilizer and a proper pH.

Further Investigations

1. What is the effect of overwatering a potted plant? Try growing seedlings in plastic cups both with and without a drainage hole. Water the plant without a drainage hole enough so that the soil surface is always covered with water.

2. Is homemade potting soil better than purchased potting soil? Try making your own potting soil by mixing different amounts of peat moss, garden soil, perlite, sand, vermiculite, and kitty litter. Then, grow seedlings in your mixtures and in a bought potting soil.

3. Will plants grow the same in all brands of potting soil? Try growing plants in several different brands of potting soil. Be sure to keep all other factors the same, including pot size, watering, light, fertilizer, and temperature.

4. Does compacting the potting soil harm plants? Compacting a potting soil occurs when you press very hard on the soil during potting. Try filling three plastic cups with the same volume of potting soil. Label the cups A, B, and C. Take the soil from cup C and pack it into cup B. You may have to push hard to get it all in. When you are finished, the level of soil in cup B should be the same as in cup A. Sow bean seeds in cups A and B, and grow them in bright light.

Further Reading

Bunt, A. C. 1988. *Media and mixes for container-grown plants*. Boston, Mass.: UnWin Hyman.

CHAPTER 15

Fertilizer Cost and Plant Growth

Dozens of brands of houseplant fertilizers are sold in drugstores, super-markets, garden centers, and department stores. They are usually in small boxes or bottles and cost $2–$3. They often have catchy names such as Miracle-Gro, Granny's Bloomers, or Jungle Juice. Which brand is best for houseplants? Which brand is the best value?

In this project, you will be a consumer scientist and determine which brand of houseplant fertilizer gives you the most fertilizer for the money. To do this, you will calculate the cost per gram of nitrogen supplied by the fertilizer. You will also study which houseplant fertilizer produces the best plant growth.

Purpose

To determine whether or not more expensive houseplant fertilizers make plants grow better.

Materials

- [ ] several brands of houseplant fertilizer
- [ ] several potted houseplants
- [ ] saucers
- [ ] marking pen
- [ ] 2-liter plastic soda bottles (see Appendix 1)
- [ ] water
- [ ] graduated cylinder
- [ ] light source
- [ ] ruler

Procedure

1. Visit various stores and purchase several brands of houseplant fertilizer. For each houseplant fertilizer, write down the brand name, the price, the guaranteed analysis, and the package net mass—for example, the brand name is Miracle-Gro, the price is $2.59, the guaranteed analysis is 15-30-15, and the package net mass is 227 g. Put these data in a table.

Brand name	Guaranteed analysis	Package mass (g)	Price ($)	Plant height (cm)
Miracle-Gro	15-30-15	227	2.59	15

2. To get the cost per gram of each fertilizer:

   a. Divide the first number in the guaranteed analysis by 100—for example, $15 \div 100 = 0.15$.

   b. Multiply the grams of fertilizer per package by the answer in step 2a—for example, $227 \text{ g} \times 0.15 = 34 \text{ g}$.

   c. Divide the package price by the answer in step 2b to get the cost per gram of nitrogen—for example, $\$2.59 \div 34 \text{ g} = \$0.076$ per gram of nitrogen.

3. Obtain several identical potted houseplants, and set each one in a saucer.

4. Label each pot with the brand name of houseplant fertilizer.

5. For liquid and powder brands, dilute houseplant fertilizer according to the package instructions in 2-liter soda bottles. A typical rate is 5 ml of fertilizer per liter of water. Water each plant with the brand of houseplant fertilizer labeled on the pot every time it is irrigated.

6. For solid brands, apply houseplant fertilizer to the potting soil in the pot labeled with that brand, and water these plants with plain water.

7. Place all the houseplants together in a warm, bright location.

8. Grow them for one month, watering as needed. Note any differences in height or appearance on the table you created in step 1.

Expected Results

The cost per gram of nitrogen will vary widely among brands of houseplant fertilizer, but there will usually be few differences in plant growth due to the brand of fertilizer.

Why?

Houseplant fertilizers come in several forms. Most are a dry powder that must be dissolved in water. Some are liquids that must be diluted with water. A few are sticks that are simply pushed into the potting soil. One type, Osmocote®, consists of soluble fertilizer coated with plastic. The fertilizer leaks out of the plastic spheres over three to four months. Osmocote is called a controlled-release fertilizer. Several houseplant fertilizers are organic, such as fish emulsion and unflavored gelatin.

All houseplant fertilizers contain nitrogen, regardless of the form or brand. Nitrogen is the mineral nutrient that plants need the most. Nitrogen is also the mineral nutrient that most often limits plant growth. Therefore, this project calculated the cost of the nitrogen in the fertilizer and did not consider other mineral nutrients.

In 1990, I did a price survey of 20 brands of houseplant fertilizer. The price of nitrogen in the fertilizer ranged from \$0.03 to \$4.66 per gram. There are many possible reasons for the wide range in price. Some fertilizers contain more of the other mineral nutrients than others. Some of the packages were more expensive than others—for example, a plastic bottle versus a cardboard box.

Probably the most important reason for the wide price range was that the concentration of the fertilizer varied. The more expensive fertilizers tended to be less concentrated. The fertilizer's guaranteed analysis is the series of three numbers that tell the percentage by mass of nitrogen, phosphorus pentoxide, and potassium oxide. The most expensive brand had 0.15% nitrogen by mass, and the cheapest brand had 21% nitrogen by mass. The fertilizer in the houseplant fertilizer package makes up very little of the total cost. Most of the price you pay goes for packaging, shipping, and handling.

Which houseplant fertilizer you use does not seem to make much difference in plant growth. One reason for this is that all houseplant fertilizers do contain nitrogen, which is the most important mineral nutrient needed by plants. Also, houseplant growth is frequently limited more by other factors than lack of fertilizer. Houseplant growth is often resticted by lack of light, lack of water, overwatering, or low humidity. Thus, a smart shopper will buy the houseplant fertilizer that has the lowest cost per gram of nitrogen.

Further Investigations

1. Can common household items be used to make a houseplant fertilizer? Try using unflavored gelatin or diluted ammonia cleaner (without soap or detergent) as a source of nitrogen fertilizer for potted plants.

2. What is the best rate of houseplant fertilizer? Try applying houseplant fertilizer at 0.5,1, and 1.5 times the recommended rate. Apply no fertilizer as a control.

3. Are organic fertilizers better for plants than inorganic fertilizers? Try fertilizing plants with an inorganic fertilizer, like Miracle-Gro, and an organic fertilizer, like fish emulsion.

Further Reading

Hershey, D. R. 1990. Sleuthing the nutrients that make your houseplant grow. *Science Activities* 27(4): 17–20.

CHAPTER 16

Soil pH and Plant Growth

Soil pH is often measured because it is a fairly easy measurement to make and because it can predict plant growth. Plants do not grow well if the soil pH is too low or too high. In this project, you will examine the effect of soil pH on plant growth and try to determine why soil pH is so important. You will also examine the effect of liming the soil on pH and plant growth.

Purpose

To determine the effect of soil pH on plant growth.

Materials

- □ sphagnum peat moss
- □ water
- □ plastic bags
- □ 4 plastic pots
- □ marking pen
- □ plastic electrical tape
- □ balance
- □ spoon
- □ powdered calcium carbonate or limestone (available at a garden center)
- □ seeds
- □ 3 saucers
- □ light source
- □ ruler
- □ 35-mm film can

☐   plastic cups
☐   distilled water
☐   pH meter or pH paper

Procedure

1. Moisten the sphagnum peat moss with hot tap water. This can be done easily by mixing the peat moss and water in a plastic bag. It should be just moist enough that no water drips out when you squeeze a handful.

2. Fill 3 of the 4 plastic pots with moist peat moss, and tap each container on a tabletop to settle it. Continue to fill the pots until the peat moss is within 1 cm of the top.

3. Label the pots A, B, and C. Pot A is the control.

4. Determine the liters of peat moss in a pot.

    a. Tape the drainage holes on the empty pot shut with electrical tape. Put tape on both sides to get a watertight seal.

    b. Measure the mass of the empty pot in kilograms.

    c. Fill the pot with water to 1 cm from the top, and measure the mass of the water-filled pot in kilograms.

    d. Subtract the mass of the empty pot from the mass of the water-filled pot. The difference is the volume of peat moss per pot in liters. Use this value to calculate the grams of calcium carbonate to add in steps 5 and 6. For example, with 0.7 liter of peat moss per pot, add 0.7×3.5 g = 2.45 g of calcium carbonate to pot B.

5. Dump pot B into a plastic bag and sprinkle in 3.5 g of calcium carbonate per liter of peat moss. Mix thoroughly, and put the peat moss back in the pot.

6. Dump pot C into a plastic bag and sprinkle in 7 g of calcium carbonate per liter of peat moss. Mix thoroughly, and put the peat moss back in the pot.

7. Sow seeds in each pot, and set pots in saucers. Water, and place in a bright, warm location.

8. Measure plant height daily for several weeks.

9. When differences are apparent, measure the pH of the peat moss in pots A, B, and C as follows:

    a. Remove the roots and loosely pack peat moss from one of the pots into a 35-mm film can so it is full and level.

b. Empty the film can full of peat moss into a clean plastic cup.

c. Add two film cans full of distilled water and mix with a clean spoon.

d. Insert the probe end of the pH meter or a piece of pH paper into the peat moss/water mixture to measure the pH. Do not use cheap pH meters with a metal probe, because they are inaccurate. Read the pH only to tenths of a unit—for example, 5.5, not 5.48.

10. Graph plant height against peat moss pH.

Expected Results

Adding limestone increases the peat moss pH, and plants generally grow better at the higher soil pH.

Why?

The pH is a measure of the concentration of hydrogen ions (H^+). A pH of 7 is neutral. A pH below 7 is acidic. A pH above 7 is basic or alkaline. There are 10 times more H^+ at pH 7 than at pH 8, and 10 times more H^+ at pH 6 than at pH 7.

The soil pH is important for plants because it affects the availability of mineral nutrients. Many mineral nutrients may be in the soil but may not be in the right form for the root to absorb them. If they are not in the right form, then they are not available to the root. The soil pH affects the form of many mineral nutrients. For example, iron is soluble and available when the pH is below 5. Iron is insoluble or unavailable when the pH is above 7. The best pH for the growth of most types of plants is between 6 and 7. Some kinds of plants grow best with a pH of 4–5, such as azaleas and blueberries.

Soil pH is increased by adding liming materials, which is called liming. The most common liming material is calcium carbonate, which makes up eggshells, seashells, and marble statues. In high rainfall areas, the soil pH tends to decrease because of the carbonic acid or other acids in the rain (see Chapter 9). Sphagnum peat moss typically has a pH of about 4, so it requires liming.

In low rainfall areas, the soil pH tends to increase. The increase in soil pH is caused by the buildup of calcium carbonate in the soil. Soil pH is decreased by adding sulfur or aluminum sulfate. Sulfur lowers the pH because bacteria convert the sulfur into sulfuric acid.

Further Investigations

1. What happens if the soil pH is too high? Try adding 25 g of calcium carbonate per liter of moist peat moss and growing plants in it. As a control, use 7 g of calcium carbonate per liter. Measure plant growth and soil pH.

2. How do you acidify soil? Try adding 25 g of calcium carbonate per liter of moist peat moss. Measure the soil pH. Add 0, 10, and 20 g of aluminum sulfate per liter of the calcium carbonate-treated peat moss. Measure the peat moss pH.

Further Reading

Hershey, D. R. 1992. Evaluating metal probe meters for soil testing. *American Biology Teacher* 54: 436–38.

CHAPTER 17

Plant Growth with Controlled-Release and Soluble Fertilizers

Osmocote is a controlled-release fertilizer, which steadily leaks fertilizer for three or four months. Do plants fertilized with Osmocote grow better than plants given a soluble fertilizer, such as Miracle-Gro? In this project, you will determine whether plants grow better with controlled-release or soluble fertilizer. You will also study how soluble and controlled-release fertilizers differ.

Purpose

To determine whether plants grow better with soluble or controlled-release fertilizer.

Materials

- ☐ Miracle-Gro fertilizer (15-30-15)
- ☐ water
- ☐ 2-liter plastic soda bottle (see Appendix 1)
- ☐ 2 plastic pots
- ☐ potting soil
- ☐ seeds
- ☐ marking pen
- ☐ 2 saucers
- ☐ light source
- ☐ Osmocote fertilizer (14-14-14)
- ☐ ruler
- ☐ 35-mm film can

☐ plastic cup
☐ distilled water
☐ spoon
☐ pH meter or pH test paper
☐ balance (optional)

Procedure

1. Dissolve Miracle-Gro fertilizer in 2 liters of water in a soda bottle. Follow the directions on the package.

2. Fill two plastic pots with moist potting soil, sow seeds, and label the pots A and B.

3. Place each pot in a saucer in a warm, bright place.

4. Water pot A with the Miracle-Gro solution as required. Keep a record of how much Miracle-Gro solution is applied.

5. Add the recommended amount of Osmocote to pot B. Water plant B with plain water as needed.

6. Observe the plants and measure the plant heights daily.

7. After four weeks, measure the pH of the soil in pots A and B as follows:

    a. Remove the roots, and loosely pack the potting soil into a 35-mm film can so it is full and level.

    b. Empty the film can full of potting soil into a clean plastic cup.

    c. Add two film cans full of distilled water and mix with a clean spoon.

    d. Insert the probe end of the pH meter or a piece of pH paper into the potting soil/water mixture to measure the pH. Do not use cheap pH meters with a metal probe, because they are inaccurate. Read the pH only to tenths of a unit—for example, 5.5, not 5.48.

8. Optional: Weigh how much dry Osmocote you added to a pot. For example, if you added a spoonful of Osmocote per pot, determine the mass of a spoonful of Osmocote. Multiply the grams of Osmocote per pot by the percentage of nitrogen in the Osmocote and divide by 100 to get the grams of nitrogen that were added. For example, 14% nitrogen × 6 g ÷ 100 = 0.84 g of nitrogen. Multiply this number by 0.25, because only about 25% of the fertilizer

should have been released in one month. For example, 0.84×0.25 = 0.21 g nitrogen supplied per pot.

9. Optional: Weigh how much dry Miracle-Gro was added per 2-liter bottle. Multiply the grams of Miracle-Gro by the percentage of nitrogen and divide by 100 to get the grams of nitrogen per 2-liter bottle. For example, 15×4 g $\div 100 = 0.6$ g of nitrogen per 2-liter bottle. Multiply this number by the number of bottles of Miracle-Gro solution applied to the pot during the experiment. For example, 0.6 g $\times$ 2 bottles = 1.2 g nitrogen applied per pot.

10. Compare the answers in steps 8 and 9 to determine whether more nitrogen was supplied by Osmocote or Miracle-Gro.

Expected Results

The plants will probably grow about equally as well.

Why?

Osmocote consists of a soluble fertilizer coated with a plastic to form little spheres. When wetted, the Osmocote spheres absorb water and swell. The fertilizer dissolves and slowly leaks out through tiny holes in the plastic coating. The rate of fertilizer leaking is affected mainly by temperature. The higher the temperature, the faster the leaking. The roots can only absorb the fertilizer that has leaked out. Miracle-Gro is a soluble fertilizer, so all the fertilizer added to the pot dissolves and can be absorbed by the roots.

Plants will usually grow equally well with controlled-release and soluble fertilizers because they both supply adequate nitrogen for the plant. The advantage of Osmocote is that it automatically fertilizes the plant. You have to remember to mix and apply soluble fertilizers, such as Miracle-Gro. Controlled-release fertilizers also reduce fertilizer leaching because less fertilizer is usually applied (see Chapter 8).

Further Investigations

1. What affects the rate of fertilizer release from Osmocote? Try placing two 10-g samples of Osmocote each in 1 liter of water. Place one jar of water in the refrigerator, and keep one at room temperature. Measure the electrical conductivity (EC) of the solution every week for four weeks (see Chapter 8). Allow the refrigerated fertilizer to

warm to room temperature just before each EC measurement. Graph the EC against time for each temperature. If you do not have an EC meter, air-dry the Osmocote samples after four weeks. Then, reweigh them to determine how much fertilizer was released.

2. Does it make a difference if the Osmocote is placed on top of the soil or mixed into the soil? Try mixing Osmocote into the potting soil before planting. Compare plants grown in that soil to plants grown in potting soil with Osmocote just on the soil surface.

3. Is leaching of fertilizer greater with Osmocote or Miracle-Gro fertilizer? Try this project and measure the EC of the leachate (see Chapter 8).

PART V

Hydroponics

The word **hydroponics** was introduced in 1937 by William Gericke, a University of California scientist. He defined hydroponics as crop production with the roots in **nutrient solution** instead of soil. A nutrient solution is a mixture of mineral nutrients and water. Growing plants in solution was not new in the 1930s. Research plants had been grown in solution culture since the mid-1800s. However, hydroponics created a sensation in the 1930s because Gericke stated that plant growth in hydroponics was much greater than with soil-grown plants. People rushed to invest in hydroponic production but soon discovered that growth of hydroponic plants was not any better than soil-grown plants. The misinformation about hydroponics was cleared up by two other University of California scientists, Dennis Hoagland and Daniel Arnon, who wrote an excellent booklet for the public in 1938.

Although hydroponics is not as miraculous as initially assumed, it remains an important research method because the roots are visible. Also, unlike in soil, the root environment in hydroponics is easily measured and changed. NASA will be using hydroponics in the space station for crop production. The space station hydroponics will also recycle carbon dioxide produced by the crew and make oxygen. Hydroponics has become an important tourist attraction at the Land exhibit in Walt Disney World's EPCOT Center in Orlando, Florida. Hydroponics is also a popular hobby, and there is limited commercial hydroponics. For example, PhytoFarm in DeKalb, Illinois, produces hydroponic lettuce under electric light in a warehouse.

Over the years, the definition of hydroponics has changed. Today, hydroponics is often defined as crop production without soil. This includes crop growth in sand, gravel, perlite, vermiculite, and other solid materials. In this book, we will consider hydroponics as originally defined: crop production with the roots in nutrient solution. For science projects, hydroponics in solution culture is best.

Hydroponic equipment for science fair projects is easily assembled from plastic soda bottles. A standard hydroponic system for science projects is described in Appendix 3. It will be used in all but one of the hydroponic projects that follow. Appendix 4 provides information on preparing seedlings and rooted cuttings for solution culture hydroponics.

FURTHER READING

Field, R. 1988. "Old MacDonald has a factory." *Discover* 9(12): 46–51.

Hershey, D. R. 1994. Solution culture hydroponics: History and inexpensive equipment. *American Biology Teacher* 56: 111–18.

Hoagland, D. R., and D. I. Arnon. 1950. *The water-culture method for growing plants without soil.* California Agricultural Experiment Station Circular 347, revised.

Murphy, W. 1984. *The future world of agriculture: Walt Disney World EPCOT Center book.* Danbury, Conn.: Grolier.

Schwartzkopf, S. H. 1992. Design of a controlled ecological life support system. *BioScience* 42: 526–35.

CHAPTER 18

Nutrient Solution Aeration and Plant Growth

A well-watered soil in the field is about 50% solids, 25% air, and 25% water, by volume, and a well-watered potting soil in a container is about 15% solids, 15% air, and 70% water (see Part IV). The air-filled pores are important because they act as air tunnels. Air from above the soil can flow into the soil and to the roots through the air-filled pores. Air contains 21% oxygen, by volume. All plant roots require oxygen because they must respire.

In hydroponics, the roots are surrounded by 100% water. Unlike soil, there are no solids or air spaces in solution culture hydroponics. How, then, do hydroponic plants get oxygen? In this project, you will determine how roots in hydroponics get oxygen.

Purpose

To determine whether or not hydroponic plants can be grown successfully without bubbling air through the solution.

Materials

- ☐ 2 hydroponic systems (see Appendix 3)
- ☐ hydroponic nutrient solution, such as Hoagland solution number 1 (see Appendix 3)
- ☐ marking pen
- ☐ rooted houseplant cuttings (see Appendix 4)
- ☐ light source
- ☐ Mariotte bottle (see Appendix 3)

☐   ruler
☐   plant press (optional)

Procedure

1. Fill two hydroponic systems with the hydroponic nutrient solution, such as Hoagland solution number 1 (Table A3.2 in Appendix 3).
2. Label the systems A and B.
3. Place rooted houseplant cuttings in each of the hydroponic systems. Be sure that the roots are all under the surface of the nutrient solution.
4. Bubble air in system A continuously with an aquarium pump.
5. Do not bubble air in system B.
6. Grow the hydroponic plants for three to four weeks under bright light and warm temperatures.
7. Keep the level of solution in both reservoirs constant by using a Mariotte bottle (see Appendix 3).
8. When plant growth differences are apparent, measure plant height.
9. If desired, dry and press the plants between newspapers under a stack of heavy books or in a plant press. The pressed plants can be part of your project report or poster.

Expected Results

Plants without air bubbling in the nutrient solution will not grow as well as plants with air bubbling.

Why?

In soil, most plant roots get oxygen from the air-filled soil pores. If the soil is overwatered, there are not enough pores filled with air. Roots in overwatered soil suffer from a lack of oxygen. Also, microbes in overwatered soil produce chemicals that are toxic to roots. Roots can suffer from lack of oxygen directly and indirectly from the toxic chemicals produced by the microbes.

Some types of plant roots grow well in overwatered soils. For example, rice thrives in soil covered with water. Rice roots get oxygen from air channels within the stems and roots. These air channels in the plant are called **aerenchyma.** The air moves from above the soil, into the aerenchyma, and to the roots.

Unlike soil, there are no air-filled pores in a hydroponic nutrient solution. Also, most types of plants do not have aerenchyma. Most hydroponic plants must get oxygen from the nutrient solution. However, water holds only a small amount of dissolved oxygen, so it is quickly consumed by the roots. To give the roots enough oxygen, air is constantly bubbled through the nutrient solution. This bubbling replenishes the oxygen supply and is called **aeration.**

Air pumps were not readily available in the 1930s. Early hydroponic growers provided oxygen to roots by keeping about half the roots above the nutrient solution. The hydroponic reservoir was kept half full. The roots above the solution remained wet because of the moist air in the top half of the reservoir.

Further Investigations

1. Will some types of plants grow well in a hydroponic system without aeration? Try growing different types of plants with and without aeration. Look at cut stem sections under a microscope for aerenchyma.

2. Will plants grow well without aeration if the reservoir of the hydroponic system is only kept half full of nutrient solution? Try growing plants with the reservoir full and half-full both with and without aeration. Use Mariotte bottles to keep the solution levels constant.

3. Is continuous aeration of the nutrient solution needed for best plant growth? Try controlling the aquarium pump with a 24-hour timer. For example, on every other hour, on 6 hours and off 6 hours, or on 12 hours and off 12 hours. Use continuous aeration as a control treatment.

4. Will adding hydrogen peroxide to the nutrient solution provide oxygen for the roots? Hydrogen peroxide (H_2O_2) is sold to clean small wounds. It bubbles on skin contact, and the bubbles are oxygen. One brand of houseplant fertilizer, Oxygen-Plus®, contains peroxide. It supposedly provides oxygen to roots of potted plants. Try adding hydrogen peroxide to both aerated and unaerated nutrient solutions. Compare the hydrogen peroxide-treated plants to plants grown with aerated and unaerated nutrient solutions lacking hydrogen peroxide.

Further Reading

Esau, K. 1977. *Anatomy of seed plants.* 2d ed. New York: John Wiley and Sons.

CHAPTER 19

Plant Growth in Various Nutrient Solutions

Preparation of the nutrient solution is one of the most difficult parts of hydroponics (see Appendix 3). Can soluble houseplant fertilizers, such as Miracle-Gro, be used to make hydroponic nutrient solutions? They are more widely available and cheaper than hydroponic fertilizers. In this project, you will determine whether or not houseplant fertilizers will make satisfactory hydroponic nutrient solutions.

Purpose

To determine whether or not houseplant fertilizers can be successfully used as hydroponic nutrient solutions.

Materials

- □ 2 hydroponic systems (see Appendix 3)
- □ marking pen
- □ water
- □ hydroponic nutrient solution, such as Hoagland solution number 1 (see Appendix 3)
- □ houseplant fertilizers, such as Miracle-Gro
- □ rooted houseplant cuttings (see Appendix 4)
- □ light source
- □ Mariotte bottle (optional)
- □ ruler
- □ plant press (optional)
- □ pH meter (optional)

112

Procedures

1. Label one hydroponic system A, and fill it with the hydroponic nutrient solution, such as Hoagland solution number 1 (Table A3.2 in Appendix 3).

2. Label the second hydroponic system B, and fill it with a houseplant fertilizer solution. Follow the package directions for mixing the houseplant fertilizer.

3. Place rooted houseplant cuttings in each of the hydroponic systems.

4. Grow plants for three to four weeks under bright light and warm temperatures.

5. Refill the reservoirs with water as needed, or use a Mariotte bottle (see Appendix 3) to automatically keep the reservoirs full.

6. When plant growth differences are apparent, measure plant height.

7. If desired, dry and press the plants between newspapers under a stack of heavy books or in a plant press.

8. If desired, measure the pH of the two hydroponic solutions.

Expected Results

Plants will grow poorly or die when grown in houseplant fertilizer solutions instead of hydroponic nutrient solutions.

Why?

Houseplant fertilizers are designed to be applied to potting soils. They do not work well in solution culture hydroponics because they are missing some of the mineral nutrients that plants need, nutrients that are already present in potting soil. Houseplant fertilizers often contain just nitrogen, phosphorus, and potassium. These are the three mineral nutrients that are usually most lacking in potting soils. A hydroponic nutrient solution must contain all 14 of the mineral nutrients for plants (see Part IV). However, you actually only have to supply 13 of these mineral nutrients—all but nickel, because nickel is present as a contaminant in the compounds containing the other 13 nutrients.

Also, many houseplant fertilizers have most of the nitrogen as ammonium rather than nitrate. Hydroponic nutrient solutions have most or all of their nitrogen as nitrate. Too much ammonium in the nutrient solution is toxic to hydroponic plants. Bacteria change ammonium into nitrate in soils. However, these bacteria often do not function as well in nutrient solutions. Also, ammonium uptake by roots and ammonium conversion into nitrate cause the nutrient solution pH to drop too low for good plant growth.

Further Investigations

1. Can you make a hydroponic nutrient solution using chemicals commonly found around the house or that you can buy at a drugstore or supermarket? Remember that you need to provide 13 mineral nutrients (see Part IV).

2. What mineral nutrients were missing from the houseplant fertilizer? Look at the houseplant fertilizer package or test the houseplant fertilizer solution using simple chemical tests (see the Further Reading section).

3. Can you add additional chemicals to a houseplant fertilizer to make it into a hydroponic nutrient solution? Try adding some of the stock solutions in Table A3.1 in Appendix 3 to the houseplant fertilizer solution. Determine whether or not plant growth improves.

4. Is one brand of hydroponic fertilizer better than others? Try growing plants with different brands of hydroponic fertilizer (see Appendix 5 for hydroponic suppliers).

Further Reading

Hershey, D. R. 1990. Pardon me, but your roots are showing. *Science Teacher* 57(2): 42–45.

Hershey, D. R., and G. W. Stutte. 1991. A laboratory exercise on semiquantitative analysis of ions in nutrient solutions. *Journal of Agronomic Education* 20: 7–10.

CHAPTER 20

Plant Mineral Nutrient Deficiency Symptoms

Plants often do not get enough of a mineral nutrient. This is called a deficiency. Plants deficient in a mineral nutrient do not grow well and usually have visible symptoms. The deficiency symptoms are often different for each mineral nutrient. For example, calcium-deficient plants have dead shoot and root tips, and iron-deficient plants have young leaves that are yellow or white.

Knowing the deficiency symptoms is useful for determining what is wrong with plants in your home, yard, or farm so that you can provide the proper fertilizer to correct the problem. In this project, you will study plant nutrient deficiency symptoms. You will learn to recognize the symptoms caused by the lack of a certain mineral nutrient.

Biological supply companies (see Appendix 5) sell kits of premeasured salts to make nutrient solutions each deficient in one mineral nutrient. If you want, you can use these instead of mixing your own solutions for steps 4 to 7.

Purpose

To see what happens when a plant does not get enough of a mineral nutrient.

Materials

☐ devil's backbone plantlets

☐ plastic container

☐ light source

☐ scissors

☐ seven 35-mm black film cans and caps

☐ paper punch

☐ marking pen

☐ stock solutions (see Appendix 3)

☐ distilled water

☐ base from a 2-liter plastic soda bottle (see Appendix 1)

☐ plastic Petri dish

☐ fluorescent light system (see Appendix 2)

☐ ruler

☐ plant press (optional)

Procedure

1. Float devil's backbone plantlets in a shallow layer of tap water in the plastic container, and place them in bright light. Roots should be well-formed in 7–10 days.

2. With the scissors, cut a slit from the edge to the center of the seven film can caps (Figure 20.1).

3. Lift the cut edges apart, and punch a hole in the center of each cap with a paper punch (Figure 20.1). The film can and cap make up a hydroponic system.

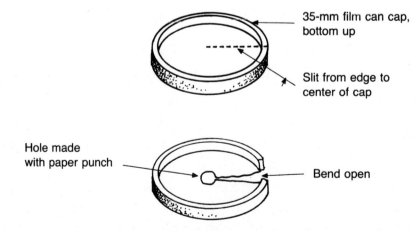

Figure 20.1. Cutting a 35-mm film can cap for use in hydroponics.

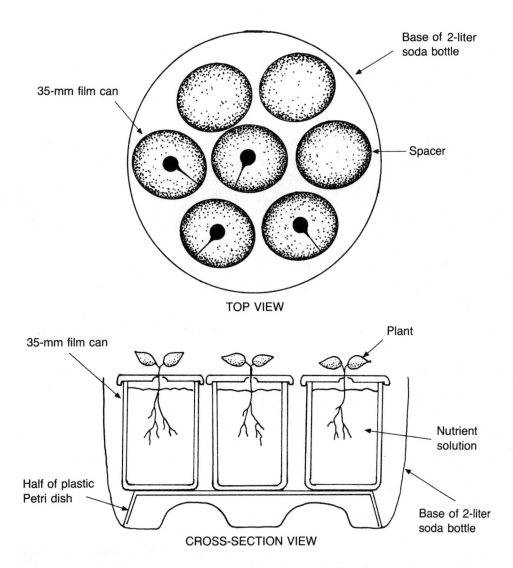

Figure 20.2. Hydroponics in a 35-mm film can held in the base of a 2-liter plastic soda bottle.

4. Label one film can hydroponic system "control," and fill it with Hoagland solution number 1.

5. Label a second film can hydroponic system "minus iron," and fill it with Hoagland solution number 1 minus the iron-EDTA stock solution. Prepare the solution using distilled water.

6. Label a third film can hydroponic system "minus phosphorus," and fill it with Hoagland solution number 1 minus the mono-potassium phosphate stock solution. Prepare the solution using distilled water.

7.   Label a fourth film can hydroponic system "minus calcium," and fill it with Hoagland solution number 1 minus the calcium nitrate stock solution. Prepare the solution using distilled water.

8.   Place a well-rooted devil's backbone plantlet in each hydroponic system (Figure 20.2).

9.   Place the film cans in the base of a 2-liter soda bottle to keep them from being knocked over. The other three film cans can be used as spacers. Put half of a Petri dish in the base of the soda bottle so the film cans sit on a level surface (Figure 20.2).

10. Grow the plantlets for three to four weeks under a fluorescent light system and warm temperatures.

11. Refill the reservoirs with distilled water as needed. Once a week should be enough because devil's backbone does not require much water.

12. When plant growth differences are apparent, measure plant height. Record any unusual symptoms.

13. If desired, dry and press the plants between newspapers under a stack of heavy books or in a plant press.

Expected Results

Plants not supplied a certain mineral nutrient will develop characteristic symptoms and will grow poorly.

Why?

Each mineral nutrient required by a plant has a unique function in the plant. For example, iron is required to form chlorophyll, phosphorus is part of **DNA,** and calcium is part of cell walls. When a plant does not get enough of a mineral nutrient, it gets deficiency symptoms. Leaves that turn yellow or white have **chlorosis.** Marginal chlorosis occurs if the yellowing is just at the edge of the leaf. Interveinal chlorosis occurs if the yellowing is just between the leaf **veins,** which are bundles of vascular tissue. **Necrosis** occurs when plant parts die.

The location of the symptoms is an important clue to the mineral nutrient that is the cause. Mineral nutrients that are easily moved within the plant move from the older to younger parts of the plant when they become deficient. Thus, symptoms appear first on the old leaves. Mineral nutrients that move easily in the plant are nitrogen, phosphorus, potassium, and magnesium.

Mineral nutrients that do not move easily from old to young parts of the plant include the remaining mineral nutrients except sulfur. Symptoms of these less mobile nutrients usually appear first on young leaves and buds. Sulfur is unusual, because symptoms may appear on both young and old parts of the plant at the same time.

Iron-deficient plants slow in growth compared to control plants, and their young leaves typically turn yellow or white. Phosphorus-deficient plants slow in growth compared to control plants, and their leaves turn a dark green, often with purplish undersides. Calcium deficiency causes the tips of roots and shoots to die. Calcium deficiency affects the plant very rapidly, so plants grow very little after being placed in a minus calcium nutrient solution.

Further Investigations

1. What do other nutrient-deficiency symptoms look like? Try preparing various nutrient solutions missing one mineral nutrient and growing plants in them. See the article by Hershey in Further Reading for recipes to prepare nutrient solutions deficient in other mineral nutrients.

2. Can deficiency symptoms occur if the nutrient solution pH is too high or too low (see Chapter 16)? Try raising the nutrient solution pH by adding 5 ml of powdered calcium carbonate or sodium bicarbonate (baking soda) per liter of nutrient solution. Try lowering the nutrient solution pH by adding sulfuric acid until the pH is 3.

Further Reading

Bergmann, W. 1992. *Nutritional disorders of plants.* New York: Gustav Fischer.

Epstein, E. 1972. *Mineral nutrition of plants: Principles and perspectives.* New York: John Wiley and Sons.

Hershey, D. R. 1994. Solution culture hydroponics: History and inexpensive equipment. *American Biology Teacher* 56: 111–18.

CHAPTER 21

Plant Changes in the Nutrient Solution pH

In Chapter 16, you learned that the soil pH had an important effect on the availability of mineral nutrients to the roots. In hydroponics, you must also be concerned about the pH of the nutrient solution for the same reasons that pH is important in soils.

In fact, the nutrient solution pH is even more important in hydroponics because it is more easily changed than the soil pH. Thus, hydroponic growers must measure the nutrient solution pH as often as several times a day. If the pH is too high or too low, it must be changed back to the ideal range.

In this project, you will see how easily the nutrient solution pH changes. You will also compare the nutrient solution pH changes for different types of plants.

Purpose

To determine whether or not different plant species change the pH of the nutrient solution in different ways.

Materials

- [ ] scissors
- [ ] heartleaf philodendron plant
- [ ] pothos or other houseplant
- [ ] 2 hydroponic systems (see Appendix 3)
- [ ] tap water
- [ ] 2 clear plastic bags or 2 bottoms of clear, 2-liter plastic soda bottles

120

☐ fluorescent light system (optional [see Appendix 2])

☐ Hoagland solution number 1 (see Appendix 3)

☐ pH meter, pH test paper, or pH test kit

☐ distilled water

☐ balance

Procedure

1. Use the scissors to take shoot cuttings from heartleaf philodendron and pothos or another houseplant. Each cutting should include a shoot tip and two fully expanded leaves.

   CAUTION: Do not get the sap from heartleaf philodendron on your skin, as it can cause a rash like poison ivy. If it does get on your skin, wash it off immediately with soap and water.

2. Place the cuttings of each houseplant species in a separate hydroponic system filled with tap water. Bubble air through the water. Cover the plants with the plastic bag or the bottom of the 2-liter soda bottle so they do not dry out (see Appendix 4).

3. Place the hydroponic system under a fluorescent light system or in a warm, bright location.

4. Wait two to three weeks until roots are well formed, then remove the plastic covering the plants.

5. Replace the water in the hydroponic systems with Hoagland solution number 1. The initial pH of the Hoagland solution number 1 should be about 5.3. Check it to be sure, and adjust it if necessary using sulfuric acid or calcium hydroxide.

6. Every day, before measuring pH, add distilled water to the hydroponics reservoirs to make up for water used by the plant.

7. Every day, measure the nutrient solution pH with a pH meter, pH test paper, or pH test kit. Do not use cheap pH meters with a metal probe, because they are inaccurate.

8. Graph the nutrient solution pH against the time in days for both types of plants.

9. Continue the experiment for about six weeks or until the difference in nutrient solution pH between the two types of plants is easy to see.

10. Determine the fresh and dry masses of the plants.

Expected Results

The pH of the nutrient solution should decrease with heartleaf philo-dendron, but the pH should increase with pothos and most other house-plants.

Why?

Plant roots absorb most mineral nutrients in the form of **ions.** Ions are particles in solution that have a charge. Positively charged ions are called **cations.** Negatively charged ions are called **anions.** Cations in nutrient solutions include calcium (Ca^{+2}), magnesium (Mg^{+2}), and potassium (K^+). Anions in nutrient solutions include nitrate (NO_3^-), sulfate (SO_4^{-2}), and phosphate ($H_2PO_4^-$).

The roots of most plants grown in Hoagland solution number 1 take up more anions than cations. That means they take up more negative than positive charges. The roots get rid of the extra negative charges by excreting hydroxyl ions (OH^-) into the nutrient solution. The OH^- cause the pH to increase because they combine with hydrogen ions (H^+) to form water (H_2O). This is why pothos increases the nutrient solution pH.

Heartleaf philodendron is unusual because its roots take up more cations than anions. That means the roots take up more positive than negative charges. The roots get rid of the extra positive charges by excreting H^+ into the nutrient solution. The H^+ causes the nutrient solution pH to decrease.

Further Investigations

1. How does pH change when the nutrient solution is missing one of the essential mineral nutrients? Repeat this project using nutrient solutions deficient in one of the mineral nutrients (see Chapter 20). The best nutrient solution to try is the iron-deficient one.

2. Hoagland solution number 1 contains all nitrogen as nitrate, which is an anion. What happens to the pH of the nutrient solution if some ammonium nitrogen is added? Ammonium (NH_4^+) is a cation. Pre-pare an ammonium sulfate stock solution using 132 g/liter. Try add-ing 1, 2, or 5 ml of this stock solution per liter of Hoagland solution number 1 and growing plants. What happens to the nutrient solu-tion pH? Do the pH changes occur without plants in the nutrient solution?

3. What happens to the nutrient solution pH if the roots have taken up all the mineral nutrients? Do this project and measure the electrical conductivity (EC) of the nutrient solution with an EC meter. EC is a measure of how many ions are in the nutrient solution. Hoagland solution number 1 has an EC of about 2.1 deciSiemens per meter (dS/m). The EC goes to 0 when all the ions have been taken up by the roots. Graph both pH and EC against the time in days.

4. What happens to the nutrient solution pH if the plants are put in the dark for several days?

Further Reading

Hershey, D. R. 1992. Plant nutrient solution pH changes. *Journal of Biological Education* 26: 107–11.

Hershey, D. R., and S. Sand. 1993. Electrical conductivity. *Science Activities* 30(1): 32–35.

Plastic Soda Bottles

One- or 2-liter plastic soda bottles are featured in many of the projects in this book. Clear bottles are generally preferred. Two-liter bottles come in two styles. The two-piece style has a removable base. Black bases are preferred because they exclude light. The one-piece style has five feet on the bottom. Bottles with plastic labels are also preferred; bottles with paper labels have a sticky glue that is hard to remove. This appendix explains how to prepare and cut plastic soda bottles to use in the projects.

PREPARING SODA BOTTLES

Materials

- ☐ 1- or 2-liter plastic soda bottle with cap
- ☐ hot tap water
- ☐ scissors
- ☐ metal lab spatula, or table knife
- ☐ paper towels (optional)
- ☐ rubbing alcohol or acetone (optional)

Procedure

1. Carefully fill the bottle about 90% full with hot tap water.
2. Cap the bottle tightly and let it stand about 1 minute.
3. If only the removable base is needed, loosen the bottle cap so that you can insert your fingers between the base and the clear part of the bottle. Gently pry the base off the bottle. The hot water will have

softened the glue holding the base on the bottle, so it should not be too hard to remove. Empty the baseless bottle and recycle it.

4. If the entire bottle is needed, gently peel off the label, cutting it with the scissors if necessary for removal.

5. Scrape off the excess glue and bits of label using the metal lab spatula or the back of a table knife.

6. If desired, remove the glue residue by rubbing with paper towels moistened with rubbing alcohol or acetone.

7. Empty water from the bottle and turn it upside down to dry the inside.

CUTTING SODA BOTTLES

Soda bottles often need to be cut to make pots, hydroponic systems, seed germinators, or funnels. You will need an adult to help you cut the bottles.

Materials

☐ 1- or 2-liter plastic soda bottle with cap
☐ marking pen
☐ ringstand with right-angle clamp
☐ ruler
☐ single-edge razor blade, or sharp-tipped knife
☐ short-bladed scissors

Procedure

1. Clamp the marking pen in the ringstand clamp (Figure A1.1).

2. Place the ringstand on a tabletop, and adjust the height of the clamp. Use the ruler to measure the height from the tabletop to the tip of the marking pen. Use the following heights:

    • funnel—23 cm.

    • hydroponic reservoir—20 cm.

    • seed germinator—15 cm.

    • pot—up to 20 cm.

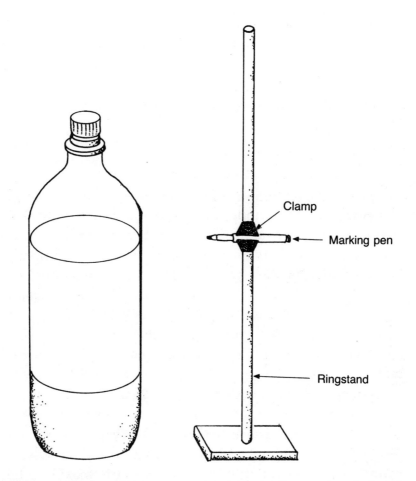

Figure A1.1. Marking a plastic soda bottle, prior to cutting, with a ringstand, clamp, and marking pen. The bottle is rotated against the marking pen.

3. Set the capped soda bottle on the tabletop. Gently rotate the base of the soda bottle while the tip of the marking pen is pressed against the soda bottle. Stop when the line completely circles the bottle.

4. With adult help, carefully puncture the bottle on the line with a single-edge razor blade or knifetip. Extend the cut to about 3 cm.

5. Complete the cut using the short-bladed scissors.

Further Reading

Ingram, M. 1993. *Bottle biology.* Dubuque, Iowa: Kendall/Hunt.

Fluorescent Light Systems

Plants can be grown under many types of electric lights but the best for science projects are fluorescent lamps because they are readily available and inexpensive. This appendix describes two fluorescent light systems you can build for your plants. One system uses straight tubes, which require special fluorescent fixtures. The other uses circular tubes, which fit a normal incandescent light bulb socket. Regardless of the type of tube, use cool-white lamps, because they are much less expensive than Gro-Lux® or other types specially designed for plant growth. Cool-white bulbs will grow plants just as well as Gro-Lux bulbs.

Generally, the lights should be on 24 hours a day, except for **photoperiod** experiments or for plants that are injured by continuous light, such as tomatoes and chia. To obtain a shorter photoperiod, you can use a 24-hour timer to automatically turn lamps on and off. Keep the top of the plants within 2–4 cm of the lamps in order to maximize the light level.

It is most efficient to hang all the lamps at the same height so light overlaps between fixtures as much as possible. If plants of various heights are grown, raise the shorter plants closer to the lamps by setting the plant pots on overturned pots or wood blocks. You will need some adult assistance—from a science teacher, industrial arts teacher, or parent—in order to assemble one of the fluorescent light systems in this appendix.

Exercise caution when using electricity. Keep water off all fluorescent light fixtures and electrical cords. Never touch the fixtures, lamps, or cords if your hands are wet. Unplug the fixtures before changing loose or burned out bulbs.

STRAIGHT TUBES

The simplest and cheapest fluorescent light source is the shop fixture with two 120-cm long, 40-watt tubes. These fixtures typically retail for under $10. For added safety, cover the sharp edges of the reflectors with electrical tape. One or two fixtures can be used if space or budget is limited. The recommendation for Wisconsin fast plants is to use three or four fixtures placed side by side to provide greater light.

Carolina Biological Supply sells a kit for a Wisconsin fast plant light bank that includes everything (fixtures, hardware, stand, cord) but the tubes (see Appendix 5). I have used the three-fixture system successfully for growing numerous other plants. The fixtures can be suspended over the plants in several ways. One of the easiest is to build a frame from 1-inch size, schedule 40 polyvinyl chloride (or PVC) pipe and fittings. PVC pipe is sized using inches. The name, however, does not correspond to the actual pipe diameter. To avoid confusion, the PVC pipe diameter is identified using the inch terminology rather than a metric measure of the actual diameter. All the materials required should be available at a hardware store.

Materials

☐ two 120-cm-long pieces of 1-inch size PVC pipe, schedule 40

☐ eight 60-cm-long pieces of 1-inch size PVC pipe, schedule 40

☐ PVC pipe cutter, or hacksaw

☐ tape measure

☐ four 1-inch size PVC three-way corner connectors

☐ four 1-inch size PVC elbow connectors

☐ 3 fluorescent shop fixtures

☐ 12 S-hooks

☐ six 45-cm pieces of chain

☐ pliers

☐ heavy-duty extension cord with three 3-prong outlets

Procedure

1. Ask the hardware store to cut the PVC pipe to the required lengths. If you cannot get the store to cut it, get an adult to help cut the pipe with a PVC pipe cutter or hacksaw.

2. Assemble the PVC pipe frame as shown in Figure A2.1. Twist the pipe firmly into the connectors. Gluing the pipe together should not be necessary.

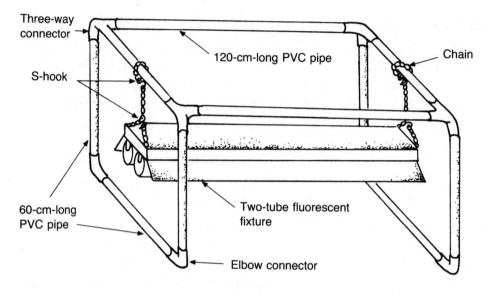

Figure A2.1. Fluorescent light system for plant growth built from PVC plastic pipe and straight-tube fluorescent fixtures.

3. Attach the small S-hooks and chain that came with the fixtures to the two holes at each end of the fixture. Bend both ends of the S-hooks closed with pliers.

4. Attach an S-hook to the bottom of each 45-cm chain. Attach the other end of this S-hook to the small chain on the light fixture. Bend both ends of the S-hooks closed.

5. Attach an S-hook to the other end of each 45-cm chain, and bend the end of the S-hook attached to the chain closed.

6. Loop the chain over the top of the PVC pipe frame. Fasten the S-hook on the chain to hold the fluorescent fixture in place (Figure A2.1). Do not close this end of the S-hook, so the fixtures can be raised or lowered.

7. When all three fluorescent fixtures are hung at the same height, plug them into the extension cord. Plug the extension cord into an electrical outlet.

CIRCULAR TUBES

The circular-tube fluorescent lamp is ideal for a miniature plant growth chamber, built with a 19-liter (5-gallon) plastic bucket. Nineteen-liter plastic buckets can often be obtained free from restaurants.

Materials

- ☐ 19-liter plastic bucket with lid
- ☐ nail
- ☐ hole saws, 7.5 and 3.8 cm diameter
- ☐ ruler
- ☐ scrap piece of lumber
- ☐ electric drill
- ☐ masking tape
- ☐ pencil
- ☐ clamp (optional)
- ☐ heavy-duty utility knife
- ☐ hacksaw blade
- ☐ rag
- ☐ 2-piece ceramic socket
- ☐ pliers
- ☐ standard extension cord
- ☐ wire stripper
- ☐ 2 wire nuts
- ☐ electrical tape
- ☐ 30-watt circular fluorescent lamp

Procedure

1. With adult help, use the nail to make five guide holes where the center of each circular hole will be. The guide hole is needed so the hole saw does not slide around on the plastic bucket. Put three guide holes in the bucket lid, one each on opposite sides of the lid about 7 cm from the edge and one in the center of the lid (Figure A2.2A). Put two guide holes 7 cm from the bottom of the bucket. Each of these last two guide holes should be directly below where the handle is attached to the bucket (Figure A2.2C).

2. With adult help, place the bucket lid on top of a scrap piece of lumber. Use the drill and a 3.8-cm hole saw to cut the hole in the center of the lid.

3. With adult help, use the 7.5-cm hole saw to cut the two holes on opposite edges of the lid for ventilation.

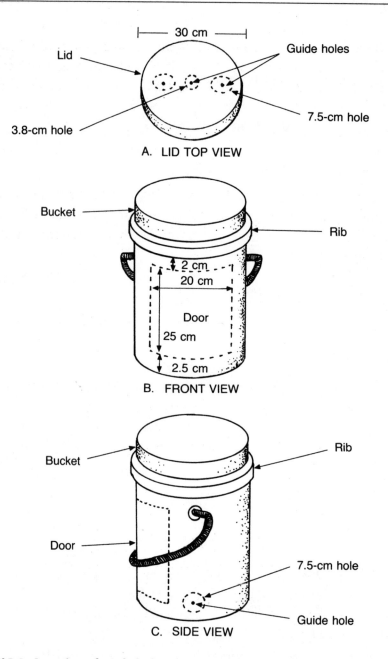

Figure A2.2. Location of guide holes, holes, and door on the 19-liter plastic bucket growth chamber.

4. With adult help, hold the bucket on its side, and cut two 7.5-cm holes on opposite sides of the bucket. The bottom edge of the hole should be about 3 cm from the bottom of the bucket (Figure A2.2C).

5. With the masking tape, ruler, and pencil, mark off a 20-cm-wide-by-25-cm-tall door on the side of the bucket (Figure A2.2B). Use the

lowest rib on the outside wall of the bucket as a guide, and make the top of the door 2 cm below it. The bottom edge of the door should be about 2.5 cm from the bottom of the bucket so there is sufficient volume to contain spills.

6. With adult help, clamp the bucket on its side or have a partner hold it firmly. Then, carefully cut through the bucket wall at one corner using the heavy-duty utility knife.

7. With adult help, cut the door out with a hacksaw blade. Insert the hacksaw blade through the cut when the cut is about 2 cm long. The blade should face away from the corner where the cut was started.

8. Wrap a rag around one end of the hacksaw blade. Use the rag as a handle and saw down to the next corner.

9. Repeat steps 7 and 8 for the other three sides of the door.

10. When the door is completely cut out, use masking tape to make a hinge on the left side or use a pair of small hinges to fasten the door to the bucket. The door should be closed when not tending the plants in order to maximize the light in the chamber.

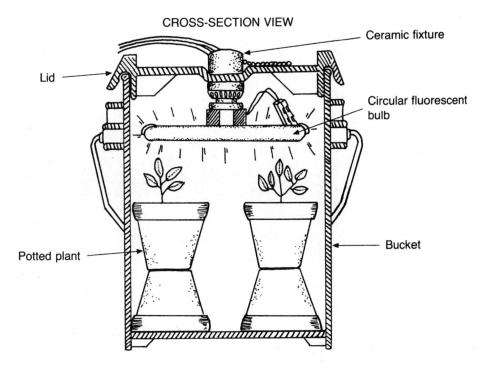

Figure A2.3. Plant growth chamber built from a plastic bucket and circular fluorescent bulb.

11. Unscrew the two-piece ceramic socket. Insert the larger piece through the top of the hole in the center of the bucket lid. Screw the small piece back onto the large piece to secure the socket in the lid (Figure A2.3).

12. With adult help, use the pliers to cut off the outlet end of the extension cord.

13. With adult help, use the wire stripper to remove 2 cm of insulation from the cut end of the extension cord.

14. With adult help, align one of the bare wires on the electrical cord with one of the bare wires on the ceramic socket. Slip a wire nut down over the two wire ends and twist. Secure the wire nut by wrapping it with electrical tape. Repeat this on the other wire.

15. Screw the circular fluorescent bulb in the ceramic socket. Plug the cord into an outlet.

Further Reading

Hershey, D. R. 1991. Plant light measurement and calculations. *American Biology Teacher* 53: 351–53.

Williams, P. H. 1991. Growbuckets and bottle reservoirs. *Wisconsin Fast Plants Notes* 4(2): 8–10.

Hydroponic Equipment and Nutrient Solutions

HYDROPONIC SYSTEM

An inexpensive hydroponic system can be prepared using 1- or 2-liter plastic soda bottles or other recyclable plastic containers (Figure A3.1). These hydroponic systems are used in Chapters 18, 19, and 21.

Materials

- ☐ materials to prepare 2-liter plastic soda bottles (see Appendix 1)
- ☐ materials to cut 2-liter plastic soda bottles (see Appendix 1)
- ☐ two 2-liter plastic soda bottles with black bases
- ☐ cork borer or electric drill
- ☐ old phone book
- ☐ aquarium tubing
- ☐ plastic drinking straw
- ☐ aquarium valve
- ☐ cotton, or polyester fiber
- ☐ aquarium pump
- ☐ aluminum foil
- ☐ nutrient solution (see "Stock Solution Preparation," later in this appendix)

Procedure

1. Prepare 2-liter soda bottles as described in Appendix 1. Remove the black base from one of the bottles.

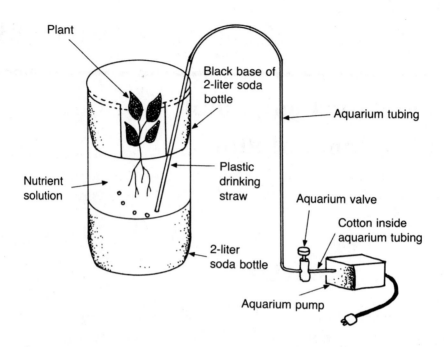

Figure A3.1. Hydroponic system built from two 2-liter plastic soda bottles.

2. Cut the bottle as described in Appendix 1. Cut the other bottle at a height of 20 cm from the bottom. The 20-cm bottle serves as the hydroponic reservoir.

3. Cut a hole in the center of the separated black base with the cork borer (Figure A3.2) or electric drill. Use an old phone book to keep the cork borer from harming your table. Cut a hole about twice the

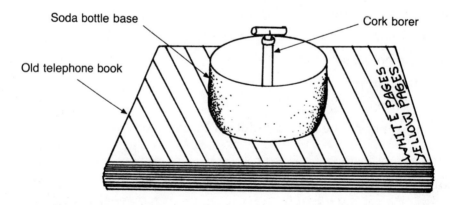

Figure A3.2. Cutting a hole in the removable base of a 2-liter soda bottle with a cork borer.

diameter of the plant stem, which allows for stem growth. This black base is the hydroponic system lid.

4. Cut a 60-cm length of aquarium tubing with the scissors. Attach one end to the aquarium valve outlet.

5. Slip a plastic drinking straw over the free end of the 60-cm piece of aquarium tubing. The plastic drinking straw keeps the tubing rigid when it is placed in the hydroponic reservoir. Slit the plastic drinking straw down one side if it will not fit over the aquarium tubing. This tubing functions as an aeration line for the hydroponic reservoir (Figure A3.1).

6. Cut a 5-cm length of aquarium tubing. Loosely pack a small wad of cotton or polyester fiber into the 5-cm section of aquarium tubing. The cotton or polyester fiber acts as a filter to prevent clogging of the aquarium valve. Attach the tubing to the aquarium valve inlet.

7. Attach the other end of the 5-cm section of aquarium tubing to the outlet of the aquarium pump.

8. Fill the hydroponic reservoir with a nutrient solution to within 2 cm of the top.

9. Insert the aeration line into the hydroponic reservoir.

10. Insert a plant through the hole in the black base (lid). Place the plant and lid on the hydroponic reservoir (see Figure A3.1). The aeration line should fit snugly between the lid and the wall of the hydroponic reservoir.

11. Plug in the aquarium pump. Adjust the aquarium valve so that air bubbles gently in the nutrient solution.

12. Wrap a 15-by-45 cm piece of aluminum foil around the clear section of the reservoir to exclude light from the nutrient solution. Fold the ends of the aluminum foil together to hold the foil in place.

MARIOTTE BOTTLE

A Mariotte bottle (Figure A3.3) is an easy method of automatically keeping a constant water level in the hydroponic reservoir. As the plant absorbs water from the hydroponic reservoir, air bubbles into the Mariotte bottle through the air-filled tube, and water flows from the Mariotte bottle into the hydroponic reservoir.

Materials

- ☐ scissors
- ☐ metric ruler
- ☐ aquarium tubing
- ☐ two-hole, #3 rubber stopper
- ☐ glass soda bottle
- ☐ 2 hydroponic reservoirs and lids (see "Hydroponic System," earlier in this appendix)

Procedure

1. Cut four pieces of aquarium tubing of the following lengths: 10, 18, 18, and 60 cm.

2. Push the two 18-cm pieces of tubing halfway into the two holes on the bottom of the two-hole, #3 rubber stopper.

3. Push the 10- and 60-cm pieces of tubing halfway into the two holes on the top of the two-hole, #3 rubber stopper. It is much easier to push two pieces of tubing into the same stopper hole from the top and the bottom than to push one piece of tubing all the way through the hole.

4. Fill a glass soda bottle with water. Insert the stopper and tubing tightly into the mouth of the bottle. The two 18-cm pieces of tubing should be inside and reach to the bottom of the bottle (Figure A3.3).

5. Set the bottle on a platform so the bottom of the bottle is the same level as the nutrient solution in the hydroponic system. This is best done by setting the bottle on the lid of an empty hydroponic system.

6. Put the free end of the 60-cm piece of aquarium tubing into the filled hydroponic reservoir (Figure A3.3).

7. Gently blow into the 10-cm piece of aquarium tubing until all the air is forced out. Air will bubble into the nutrient solution until the air is forced out. Water will automatically flow from the Mariotte bottle into the reservoir whenever water is lost from the reservoir. Air bubbles will exit the air-filled tube in the Mariotte bottle as water flows into the hydroponic reservoir (Figure A3.3).

8. Repeat steps 4 to 7 whenever the Mariotte bottle is empty. Use two or more Mariotte bottles per reservoir if desired.

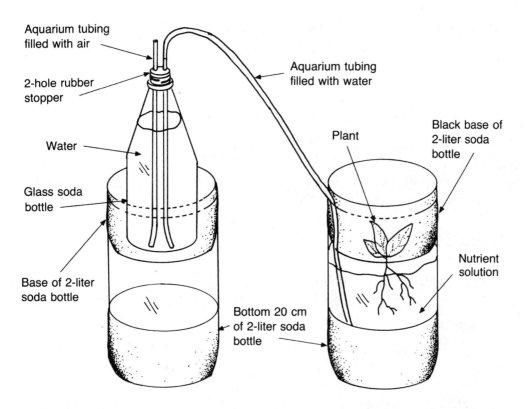

Figure A3.3. Maintaining a constant solution level in a hydroponic system with a Mariotte bottle.

STOCK SOLUTION PREPARATION

The hardest part of hydroponics is the preparation of the nutrient solution. Ask your science or chemistry teacher for help if you mix your own solutions. Table A3.1 gives amounts of each material needed for a stock solution, and preparation instructions are given below. Stock solutions are very concentrated solutions that are diluted to prepare the nutrient solution, which is the solution supplied to the plant. Table A3.2 shows how to dilute stock solutions to make a nutrient solution.

The stock solutions in Table A3.1 will be used to prepare a modified Hoagland solution number 1, which is a popular nutrient solution used by scientists since the 1930s. It has been slightly modified from the original recipe by adding iron-EDTA (iron ethylenediamine tetraacetic acid) in place of the iron tartrate in the original recipe. Iron-EDTA is a better source of iron for plants than iron tartrate.

Hydroponic solutions can be purchased from biology supply companies or from hydroponic suppliers, who sell salt mixtures that are dissolved to produce a hydroponic nutrient solution (see Appendix 5).

Materials

- ☐ top-loading balance
- ☐ analytical balance
- ☐ six 1-liter beakers
- ☐ calcium nitrate
- ☐ potassium nitrate
- ☐ magnesium nitrate
- ☐ monopotassium phosphate
- ☐ iron-EDTA
- ☐ boric acid
- ☐ manganese chloride
- ☐ zinc sulfate
- ☐ copper sulfate
- ☐ molybdic acid
- ☐ distilled water
- ☐ six 1-liter volumetric flasks
- ☐ six stoppers to fit volumetric flasks
- ☐ six 1- or 2-liter plastic soda bottles (see Appendix 1)

Procedure

1. For each solution A to E in Table A3.1, weigh out the chemical required into a separate 1-liter beaker on a top-loading balance.
2. For solution F, weigh out each of the five chemicals required on an analytical balance. Place the five chemicals in a 1-liter beaker.
3. To the 1-liter beakers in steps 1 and 2, add about 600 ml of distilled water. Mix until dissolved.
4. Pour each solution from step 3 into a separate clean 1-liter volumetric flask.
5. Fill each flask to the 1-liter mark with distilled water.
6. Stopper each volumetric flask and shake.

TABLE A3.1. Stock Solution Preparation for Modified Hoagland Solution Number 1

Stock solution	Chemical formula	Grams of chemical per liter of stock solution
A. Calcium nitrate	$Ca(NO_3)_2 \cdot 4H_2O$	236
B. Potassium nitrate	KNO_3	101
C. Magnesium sulfate	$MgSO_4 \cdot 7H_2O$	246
D. Monopotassium phosphate	KH_2PO_4	136
E. Iron-EDTA	FeNa-EDTA	18.4
F. Micronutrients		
Boric acid	$B(OH)_3$	2.86
Manganese chloride	$MnCl_2 \cdot 4H_2O$	1.81
Zinc sulfate	$ZnSO_4 \cdot 7H_2O$	0.22
Copper sulfate	$CuSO_4 \cdot 5H_2O$	0.08
85% Molybdic acid	$H_2MoO_4 \cdot H_2O$	0.02

7. Pour the stock solution from each flask into a labeled bottle. Cap tightly. Stock solutions will last for years if stored in the dark at room temperature.

NUTRIENT SOLUTION PREPARATION

Follow the directions on the package to make a nutrient solution using a packaged hydroponic fertilizer or use the stock solutions described in Table A3.1 to mix your own nutrient solutions. Nutrient solution preparation is relatively simple if stock solutions are already prepared. Table A3.2 gives the amounts of each stock solution needed for a nutrient solution. Use a separate, labeled beaker and labeled measuring container for each stock solution. This prevents accidentally contaminating one stock solution with another. Use a 19-liter (5-gallon) plastic bucket and graduated cylinders to prepare a 10-liter batch of nutrient solution. Use a 2-liter plastic soda bottle and plastic syringes to prepare a 2-liter batch of nutrient solution.

Materials

☐ 2-liter plastic soda bottle

☐ 19-liter (5-gallon) plastic bucket (optional)

- ☐ water
- ☐ masking tape
- ☐ marking pen
- ☐ 6 beakers or plastic cups
- ☐ six 5-ml plastic syringes, or 10-, 20-, and 50-ml graduated cylinders
- ☐ stock solutions (see "Stock Solution Preparation," earlier in this appendix)

Procedure for 2-Liter Batch

1. Fill a clean 2-liter soda bottle half full of tap or distilled water.

2. Use masking tape and the marking pen to label each clean beaker with the letter of each of the six stock solutions in Table A3.2.

3. Use masking tape and the marking pen to label each of six 5-ml plastic syringes with the letters of the six stock solutions in Table A3.2.

4. Pour a small volume of each stock solution into the corresponding labeled beaker.

5. Use the 5-ml syringe labeled A to draw up 5-ml of stock solution A from beaker A. Expel the solution into the 2-liter soda bottle. Repeat so a total of 10 ml of stock solution A is added to the 2-liter bottle.

6. Repeat step 5 for syringes B to F and beakers B to F. With each lettered syringe, draw up the amount of stock solution in Table A3.2.

7. Fill the 2-liter soda bottle with water, cap tightly, and mix.

TABLE A3.2. Preparation of Modified Hoagland
Solution Number 1

| | Milliliters of stock solution per batch | |
Stock solution	2-Liter batch	10-Liter batch
A. Calcium nitrate	10	50
B. Potassium nitrate	10	50
C. Magnesium sulfate	4	20
D. Monopotassium phosphate	2	10
E. Iron-EDTA	2	10
F. Micronutrients	2	10

Procedure for 10-Liter Batch

1. Fill a clean 19-liter plastic bucket with 6 liters of distilled or tap water. Measure the water with a 2-liter soda bottle or 1- or 2-liter graduated cylinder.

2. Use masking tape and the marking pen to label each clean beaker with the letter of each of the six stock solutions in Table A3.2.

3. Use masking tape and the marking pen to label two 50-ml graduated cylinders A and B. Label a 25-ml graduated cylinder C. Label three 10-ml graduated cylinders D, E, and F.

4. Pour a small volume of each stock solution into the corresponding labeled beaker.

5. Pour 50 ml of stock solution from beaker A into graduated cylinder A. Pour the solution from cylinder A into the 19-liter plastic bucket.

6. Repeat step 5 for beakers B to F and graduated cylinders B to F. For each lettered graduated cylinder add the amount of stock solution in Table A3.2.

7. Add 4 liters of water to the 19-liter bucket and stir using a 50-ml graduated cylinder as a stirring rod.

Seedlings and Rooted Cuttings for Hydroponics

SEEDLINGS

To grow seedlings for solution culture hydroponics, germinate seeds in perlite or clean, coarse sand. These materials are used because they are fairly easy to remove without damaging the roots. When seedlings are

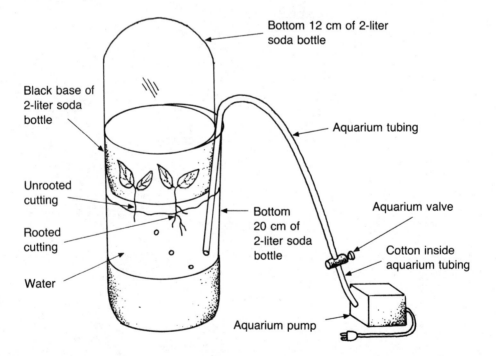

Figure A4.1. System for rooting cuttings in hydroponics built from 2-liter plastic soda bottles.

large enough, gently dig out the roots with your finger and rinse off the perlite or sand in room-temperature water. Then, place the seedlings in a hydroponic system (see Figure A3.1 or Figure 20.2). To assure that no particles stick to the roots, small seeds can also be germinated in a seed germinator built from a 2-liter soda bottle (see Figure 2.2).

ROOTED CUTTINGS

Rooted cuttings of houseplants are excellent for hydroponics. Cuttings of pothos, heartleaf philodendron, wandering jew, piggyback plant, and many other houseplants root easily without a rooting hormone. Coleus, euonymus, and geranium benefit from a rooting hormone.

Materials

☐ shoot cuttings from houseplants
☐ rooting hormone
☐ 2 hydroponic systems, one filled with tap water (see Appendix 3)
☐ clear plastic bag, or bottom 12 cm of a clear, 2-liter plastic soda bottle

Procedure

1. Remove the leaves on the lower third of the stem on the shoot cuttings.
2. Dip the cut ends of shoot cuttings in a rooting hormone.
3. Place the cuttings in a hydroponic system filled with tap water.
4. Place a clear plastic bag or the bottom 12 cm of a clear, 2-liter plastic soda bottle over the hydroponic system to hold in moisture (Figure A4.1). This prevents the cuttings from drying out.
5. Transfer the rooted cuttings to another hydroponic system when roots are well formed.

Suppliers

RETAIL STORES

Many items can be obtained locally at department stores and specialty stores. Supermarkets sell dry bean and popcorn seeds, plastic cups, paper towels, and distilled water. Drugstores sell tincture of iodine, boric acid, and epsom salts (magnesium sulfate). Pet stores sell aquarium tubing, aquarium pumps, and aquarium valves. Garden centers sell potting soils, fertilizers, pots, saucers, liming materials, and flower and vegetable seeds. Hardware stores sell equipment for fluorescent light systems (see Appendix 2) and denatured alcohol. Large department stores often sell most of the items mentioned above.

BIOLOGICAL SUPPLY COMPANIES

Your science teacher should have catalogs from one or more biological supply companies. Often it is easier for your science teacher to order supplies for you as many items are sold only to schools. The National Science Teachers Association publishes an annual directory of *Science Education Suppliers,* which lists dozens of suppliers. Major suppliers have virtually anything required for a plant science project. Carolina Biological Supply is the official marketer of Wisconsin fast plants. The addresses of this and another major supplier follow:

Carolina Biological Supply
2700 York Road
Burlington, NC 27215-3398
(800) 334-5551

146

Ward's Natural Science Establishment, Inc.
5100 West Henrietta Road
P.O. Box 92912
Rochester, NY 14692-9012
(800) 962-2660

SEEDS

Gardening magazines, such as *Horticulture, Fine Gardening,* and *Better Homes and Gardens* have ads for many mail-order seed companies. However, these companies often sell seeds in too small a quantity for science project use. The following supplier is the most inexpensive source I have found for larger amounts of flower and vegetable seeds:

R. H. Shumway
HPS Division
P.O. Box 1
Graniteville, SC 29829
(800) 322-7288

Chia seeds can be obtained from the following companies:

Native Seeds/SEARCH
2509 North Campbell Avenue, #325
Tucson, AZ 85719

Joseph Enterprises, Inc.
425 California St., Suite 1300
San Francisco, CA 94104
(415) 397-6992

Total Kaos
7901 Canoga Avenue, Unit F
Canoga Park, CA 91304

Legume and other tree seeds can be obtained from the following:

F. W. Schumacher Co., Inc.
36 Spring Hill Road
Sandwich, MA 02563-1023
(508) 888-0659

HYDROPONIC SUPPLIERS

Garden centers often do not sell hydroponic nutrient solutions, but they are available from many mail-order suppliers including the following:

Eco Enterprises
1240 NE 175th Street, #B
Seattle, WA 98155
(800) 426-6937

New Earth Indoor/Outdoor Garden Center
3623 East Highway 44
Shepherdsville, KY 40165
(800) 462-5953

Worm's Way
3151 South Highway 446
Bloomington, IN 47401-9111
(800) 274-9676

Projects to Avoid

Many plant biology science projects have been in wide use since the late 1800s. As science has advanced, some of these classic projects have been replaced by better projects, which use new materials or techniques. As scientific knowledge has expanded, scientists have discovered that other classic projects do not really show what they were supposed to show. The projects described in this book have been carefully checked for accuracy and use new materials and techniques. Many outdated or inaccurate science projects still are reprinted in other recent science project books and articles, so you need to be aware of projects to avoid.

COLLECTIONS AND MODELS

For the 12-years-and-up age range of this book, a science project with plants usually needs to include an experiment. That means you need at least two treatments with one as a control. Therefore, collections of plants or plant parts and models of plants do not usually qualify. Collections can sometimes be made the basis of experimental projects. For example, the stomatal densities and distributions could be compared among many related species by using a leaf collection (see Chapter 6). Models can sometimes be used in experiments. For example, balloons can be used as models of guard cells in determining how a stomate opens (see Chapter 6).

ART AND NATURE CRAFT PROJECTS

Plant projects may involve drying flowers, making dyes from plants, preparing food from wild plants, bark rubbings, leaf prints, plaster casts of plant parts, etc. These activities are very interesting, but they are not really science projects. They are really art or nature craft projects and should be avoided in competitions.

COUNTING TREE RINGS

A common project involves counting tree rings to determine the age of the tree. This is a valid scientific technique but it is not an experiment. There is no control treatment, and you have no way of estimating the accuracy of the age estimate. You would need to do something more to have a good project. For example, you might compare the annual rainfall with the width of each annual ring. This would require that you know the tree location, the year that the tree was cut down, and the annual rainfall for several years.

HYDROTROPISM

Many science project books have experiments on root **hydrotropism.** However, these experiments really do not demonstrate hydrotropism. They merely show that roots require water to grow.

ALLELOPATHY

Allelopathy is a popular science project. Allelopathy is an inhibitory effect of a chemical from one plant on a neighboring plant. A typical experiment involves grinding up one type of plant and spraying it on another plant. The problem with such a treatment is that it is very artificial. Plants do not naturally grind themselves up and spray themselves on neighboring plants! In a natural situation, there may not be any allelopathy, yet you may be able to "show" allelopathy when you do the artificial grinding experiment. If you do an allelopathy experiment, be very careful how you interpret the results.

HELMONT'S WILLOW TREE EXPERIMENT

One of the earliest experiments in modern biology was published in 1648 by Jean-Baptista van Helmont, a Belgian doctor, and involved growing a willow tree in 91 kg of soil. The soil was weighed again after 5 years. Helmont found that the soil mass had not changed but the willow had gained 75 kg. He concluded that the increase in the mass of the willow had come from the water the plant received. This experiment is sometimes repeated on a smaller scale as a science project.

It is not a good experiment today, however, because Helmont did not know about the basics of photosynthesis. We now know that the increase in plant dry mass is mainly due to carbon dioxide (see Part III). Therefore, Helmont's experiment is really misleading because it did not consider carbon dioxide. Also, Helmont's experiment is inaccurate because it is impossible to completely separate roots from soil.

PSEUDOSCIENCE

Science projects sometimes examine the effect of talking, type of music, sign of the moon, or extrasensory perception (ESP) on plant growth. Such projects fall into the category of **pseudoscience,** which is something that sounds scientific but is not based on real science. Astrology, or horoscopes, is another common example of pseudoscience.

Plants do not have a nervous system or brain like animals. Thus, there is no reason to expect that they will grow better if you talk nicely to them or play a certain kind of music. Plant pseudoscience experiments might offend scientists who judge your projects. This may embarrass both you and your science teacher. There are so many real science projects with plants to choose from that you should not waste your time on pseudoscience.

FURTHER READING

Gabriel, M. L., and S. Fogel. 1964. *Great experiments in biology.* Englewood Cliffs. N.J.: Prentice-Hall.

Galston, A. W. 1974. The unscientific method. *Natural History* 83(3): 18, 21, 24.

Hershey, D. R. 1991. Digging deeper into Helmont's famous willow tree experiment. *American Biology Teacher* 53: 458–60.

Hershey, D. R. 1992. Is hydrotropism all wet? *Science Activities* 29(2): 20–24.

Hershey, D. R. 1994. Allelopathy misconceptions. *American Biology Teacher* 56: 261–62.

Glossary

acid precipitation Rain, snow, sleet, or hail with a pH less than 5.6. The low pH is caused by sulfuric and nitric acids that come from the burning of fossil fuels containing sulfur and nitrogen.

acid rain Acid precipitation in liquid form.

adhesion Attraction between different types of molecules.

adventitious roots Roots that form on stems or leaves instead of on other roots.

aeration The bubbling of air in a hydroponic nutrient solution.

aerenchyma Air channels within the stem and root of some types of plants. The channels allow air to flow from the shoots to the roots.

allelopathy An inhibitory effect of a chemical from one plant on a neighboring plant.

angiosperm A flowering plant.

anion A negatively charged ion.

anther The part of the flower that produces pollen.

antitranspirant A chemical spray that closes or blocks stomates.

autotroph An organism that gets its energy from light or from inorganic molecules. Most photosynthetic plants are autotrophic.

C-3 photosynthesis A form of photosynthesis that first produces a three-carbon compound.

C-4 photosynthesis A form of photosynthesis that first produces a four-carbon compound.

carbohydrate An organic compound containing carbon, hydrogen, and oxygen in about a 1:2:1 ratio.

cation A positively charged ion.

cellulose A carbohydrate that makes up a large part of the plant cell walls.

chlorophyll   A green pigment that absorbs light energy and is needed for photosynthesis.

chloroplasts   Small structures in plant cells that contain chlorophyll.

chlorosis   An abnormal yellow or white color in a leaf that is normally green.

control   An experimental treatment that contains a zero or normal level.

cotyledon   The seed leaf or leaves that usually store energy for the embryo.

crassulacean acid metabolism   A form of photosynthesis in which plants open their stomates at night and fix carbon dioxide into organic acids. In the light, the stomates are closed and photosynthesis occurs using the carbon dioxide from the organic acids.

cross-pollination   The transfer of pollen from one plant to the stigma of another.

cultivar   A cultivated variety of a plant.

cuticle   A waxy layer on the outside of the plant epidermis.

dicot   An angiosperm with two cotyledons in the seed. Examples include radish, lettuce, and tomato.

dicotyledon   *See* dicot.

DNA   DNA is an abbreviation for deoxyribonucleic acid. The large molecules in a living organism that contain the directions for growth and reproduction are DNA molecules.

electrical conductivity   A measure of the electric current a solution will carry that is roughly proportional to the concentration of ions in the solution.

embryo   The undeveloped plant in a seed.

endosperm   A nutritive tissue in some types of seeds.

epicotyl   The part of the stem in the embryo or seedling above the cotyledon(s).

factorial   An experiment that uses two or more levels of two or more different factors. For example, 2 fertilizer levels and 2 light levels would be a 2-by-2 factorial, which would have $2 \times 2 = 4$ treatments.

fruit   A ripened ovary. Found only in angiosperms.

genus   The first word in the species name. A group of similar species.

germinate   To sprout from a seed into a seedling.

germination percentage   The number of seeds that germinate in every 100 seeds.

global warming   The expected increase in the earth's temperature caused by the increasing level of carbon dioxide and certain other gases.

glucose   A carbohydrate with six carbons that is joined together in long chains to make starch and cellulose.

guaranteed analysis   The percentages by mass of nitrogen, phosphorus pentoxide, and potassium oxide that appear on a fertilizer label.

guard cells   A specialized pair of cells in the leaf surface that surround the stomate. The stomate opens when the guard cells swell.

guttation   The flow of xylem sap out of modified stomates called hydathodes.

gymnosperm   A seed plant that does not flower. Examples include conifers, cycads, and ginkgo.

hand-pollination   Human transfer of pollen from the anther to the stigma.

heterotroph   An organism that gets its energy by consuming other organisms.

hilum   A small scar on the seed coat, which marks the place where it was attached to the parent plant.

hydathode   A modified stomate that cannot close and functions in guttation.

hydration   The uptake of water by dry seeds or other dry materials.

hydroponics   Crop production with plant roots in nutrient solution instead of soil.

hydrotropism   An old theory that roots bend toward a source of water. Experiments that attempt to show hydrotropism usually just show that roots need water to grow.

hypocotyl   The part of an embryo or seedling below the cotyledon(s) and above the radicle. The hypocotyl forms the stem below the cotyledon(s) in many types of seedlings; however, in many other types of seedlings, the hypocotyl remains below ground.

hypothesis   An educated guess of what the results of an experiment will be.

imbibition   The uptake of water by dry seeds or other dry materials.

ions   Atoms or molecules that have a positive or negative charge.

leachate   The water and dissolved substances that are leached.

leaching   Movement of water and its dissolved substances through the root zone.

leaching fraction   The volume of leachate divided by the volume of solution applied to the soil.

macronutrient   An element required by the plant in a large amount for normal growth.

micronutrient   An element required by the plant in a small amount for normal growth.

micropyle   A small opening in the seed near the hilum where the pollen tube entered.

monocot   An angiosperm with only one cotyledon in the seed. Examples include grasses, palms, tulips, and lilies.

monocotyledon   *See* monocot.

necrosis   Death of plant tissue.

nutrient solution   A mixture of water and the mineral nutrients required by plants.

ovary   The part of the flower that can develop into a fruit.

perlite   Particles of a white volcanic rock that are used in potting soils.

pH   A scale indicating the concentration of hydrogen ions (H^+) in a solution. A pH of 7 is neutral. A pH less than 7 is acidic. A pH above 7 is basic. A decrease of 1 pH unit is a tenfold increase in the concentration of H^+.

phloem   The plant tissue that moves carbohydrates and other molecules from leaves to roots and throughout the plant.

photoperiod   The number of hours of light in a 24-hour period.

photoperiodism   The effect of the photoperiod on plant growth responses, such as flowering and bulb formation.

photosynthesis   The complex process whereby chlorophyll-producing cells absorb light energy and produce carbohydrates and oxygen from water and carbon dioxide.

phototropism   The bending of a leaf, stem, root, or flower toward or away from a light source.

plant agriculture   The study of how to grow plants.

plant ecology   The study of a plant's relationships with other organisms and with its environment.

plant physiology   The study of how plants function.

plumule   The part of a plant embryo that will form the plant shoot.

pollen   The dustlike grains produced by the anther of the flower.

pollination   The transfer of pollen to the stigma of a flower.

polymer   A large molecule made by linking together hundreds or thousands of a small molecule into long chains.

potting soil   A mixture of materials, such as perlite and peat moss, used to grow plants in pots.

pseudoscience   A belief that seems scientific but is not supported by real scientific experiments.

pure seed   The type of seed listed on a seed package label.

radicle   The part of a plant embryo that will form the primary plant root.

relative humidity   The ratio of the amount of water vapor the air is holding to the total amount it could possibly hold at that temperature, expressed as a percentage.

replication   An experimental unit.

respiration   The process whereby an organism obtains energy by breaking apart organic molecules, such as carbohydrates, fats, and proteins.

root   To form adventitious roots on detached stems, leaves, or shoots.

root zone   The area in the soil where roots grow.

rosette   A plant whose stem is very short and whose leaves form a compact cluster, such as piggyback plant or dandelion.

seed coat   The outer covering of a seed.

seed purity   The percentage of pure seeds, by mass, in a sample of seeds.

species   The scientific name of an organism consisting of the genus and specific epithet—for example, *Zea mays*.

specific epithet   The second word in a scientific name. For example, *mays* is the specific epithet in the species *Zea mays*.

sphagnum peat moss   The fibrous, partly decomposed remains of moss plants, widely used in potting soils and to improve garden soils.

starch   An insoluble carbohydrate that is produced in many leaves during photosynthesis. It is also found in stems, such as potato tubers and roots.

stigma   The flower part on which pollen germinates.

stoma   *See* stomate.

stomatal density   The number of stomates per unit area of leaf surface.

stomate   An opening in a leaf that allows carbon dioxide to enter the leaf for photosynthesis. The stomate is surrounded by two guard cells.

time-course graph   A plant measurement, such as stem length or flower size, plotted on the vertical axis against time on the horizontal axis.

translocation   Movement of water, mineral nutrients, and organic compounds from place to place within the plant. It occurs in xylem and phloem.

transpiration   Loss of water vapor from plant shoots. Most transpiration occurs through the stomates.

transpiration ratio   The grams of water transpired for each gram of plant dry matter produced.

turgor pressure   The pressure inside a plant cell caused by water uptake that keeps the cell firm.

variegated   Leaf or plant that has more than one color on each leaf—for example, coleus.

vascular tissue   Plant tissue functioning in translocation and consisting mainly of the xylem and phloem.

vermiculite   A mineral material used in potting soil.

vein   A bundle of vascular tissue in a leaf.

xylem   The plant tissue that moves water and mineral nutrients from the roots to the shoots.

Index

A

B

C